Soy Protein and
National Food Policy

About the Book and Editor

Over the last twenty-five years, a healthy international business has developed in isolated soy protein food ingredients. Today, isolated soy proteins are used primarily as ingredients in processed meat products. They are also used as valuable sources of protein in medical nutritional products and in combination with dairy products around the world.

This book puts into perspective the importance of isolated soy protein food ingredients and their implications for public policy. The contributors examine increasing world demand for meat and meat products and provide case studies of Sweden, China, and Mexico that demonstrate the potential economic value of using isolated soy protein in a macroeconomic context. The contributors also provide methodology for quantifying the profit a country can realize by using isolated soy protein ingredients; show why these ingredients have achieved superiority over other vegetable and "novel" protein sources; and discuss the importance of traditional food product quality in gaining consumer acceptance.

F. H. Schwarz is senior vice president of the Agribusiness Corporation of America in Washington, D.C.

Published in cooperation with
Protein Technologies International,
a subsidiary of the Ralston Purina Company

Soy Protein and National Food Policy

EDITED BY

F. H. Schwarz

Routledge
Taylor & Francis Group

NEW YORK AND LONDON

First published in paperback 2024

First published 1988 by Westview Press, Inc.

Published 2019 by Routledge
605 Third Avenue, New York, NY 10158

and by Routledge
4 Park Square, Milton Park, Abingdon, Oxon OX14 4RN

Routledge is an imprint of the Taylor & Francis Group, an informa business

Copyright © 1988, 2019, 2024 by Protein Technologies International, Inc.

Library of Congress Cataloging-in-Publication Data
Soy protein and national food policy / edited by F. H. Schwarz.
    p. cm.
  ISBN 0-8133-7632-7
  1. Soy products. 2. Plant proteins as food. 3. Food supply—
Government policy. I. Schwarz, F. H.
TP438.S6S69 1988
338.1'7334—dc19   88-3787
                                                                       CIP

Publisher's Note
The publisher has gone to great lengths to ensure the quality of this reprint but points out that some imperfections in the original copies may be apparent.

ISBN 13: 978-0-367-28847-1 (hbk)
ISBN 13: 978-0-367-30393-8 (pbk)
ISBN 13: 978-0-429-30722-5 (ebk)

DOI: 10.1201/9780429307225

Contents

Foreword

The purpose of this book is to put into perspective the value of soy protein food ingredients and the implications of this value for public policy. The perspective includes considerations of economic value, financial profitability, technology, and consumer acceptance.

Over the last twenty-five years, a very viable international business has developed in soy protein food ingredients, especially isolated soy proteins. These ingredients find their principal use today in processed meat products.

Soy proteins have also found some uses in fortifying grains—especially for malnourished segments of the population. However, fortifying grain products causes the product cost to increase, limiting the business opportunities. Few governments have the funds available to subsidize increased costs. There is also an emerging interest in using isolated soy protein products in combination with dairy products, although dairy subsidies often distort the real economic benefits that can be provided.

Vegetable protein in general and soy proteins in particular were promoted as the answer to malnourishment. Today it is clear that soy proteins' role, especially isolated soy proteins' role, is to mitigate the cost pressure of producing processed meat products.

This business in meat products has occurred for several important reasons. First, as incomes rise, there is an increasing demand for meat and meat products. Second, using isolated soy proteins in meat products lowers their cost and therefore has the potential of lowering their price to the consumer. Third, the meat products utilizing isolated soy proteins have achieved outstanding consumer acceptance as they maintain or improve their traditional sensorial qualities. Fourth, isolated soy protein ingredients and the technology they represent provide both financial value to meat processors and economic value to the countries in which they are used.

The chapters that follow carry out the stated purpose of this volume. Chapter 1, by D. Gale Johnson, presents the issues and conclusions in a policy context. Chapter 2, by J. S. Sarma, describes the increasing demand for meat and meat products and the resulting demand for feedgrains.

Chapters 3, 4, and 5, case studies of Sweden, China, and Mexico by M. Smith, T. Sicular, and G. E. Schuh, present the potential economic value of using isolated soy protein in a macroeconomic context. Chapter 6, by F. H. Schwarz, provides the methodology for quantifying both the financial profitability and the economic profitability to the country in which these isolated soy protein ingredients are used.

Chapter 7, by W. Wolf, shows why soy protein ingredients have achieved technical superiority to other vegetable and "novel" protein sources. Chapter 8, by S. G. Sellers, describes the critical importance of traditional food product quality in achieving consumer acceptance—without which the financial and economic benefits are not viable. This is particularly important as the intense focus on human needs has often blurred the importance of food acceptance.

It is with deep appreciation that I thank the authors for their effort, insights, and contributions to this volume. We have not always agreed on each point but, as sponsors of this volume and the world's largest producer of isolated soy protein, we are deeply indebted for the careful consideration given in the writing of each chapter. Special thanks goes to Gale Johnson, who chaired the conference at which these papers were presented and discussed, and to Walter Falcon for his many helpful suggestions in guiding the entire project.

If this volume is to contribute to national food policies around the world, it will greatly enhance the international trade in soy proteins. What is needed is simply lower tariffs and greater availability of import licenses and foreign exchange allocations.

The authors have demonstrated the potential value of soy protein food ingredients. Now it is up to governments to permit freer trade so that these food ingredients can earn their markets based on the financial and economic value they offer consumers, food processors, and nations alike.

P. H. Hatfield
President
Protein Technologies International,
a subsidiary of the Ralston Purina Company

Chapter 1

Policy Implications

*D. Gale Johnson**

* D. Gale Johnson is Professor of Economics, University of Chicago.

I am not one who has believed or now believes that the world's demand for food will outrun its supply of food during the 20th Century. Thus I do not believe that for the foreseeable future international prices for the major sources of calories, such as grains, vegetable oils and sugar, will increase if such prices are adjusted for inflation. Yet I do support the development of soy protein food ingredients, including isolated soy proteins, and the fullest exploration of their economic feasibility.

Contrary to some widely held views, there has been general improvement in the nutrition and health status of the population of the developing countries. The improvement has not been uniform and it is possible that in some parts of Africa there has been deterioration in the food situation. This has certainly been the case where civil wars and insurrections have prevailed such as in Ethiopia, Sudan and Uganda. But for the area of the world for which there was great concern only a decade ago, namely Asia, there has been significant improvements in nutrition and health.

Table 1 includes data on food production per capita in the major regions of developing countries. In all regions, except Africa, there has been increases in per capita food production, modest in the near East and substantial in the Far East.

But the food available depends upon more than production; trade in food affects how much food is supplied to the domestic market. In the early 1960s Africa was a net exporter of food; at the end of the 1970s Africa was a major importer of food. In 1979-81 Africa imported 16 percent of the calories that it consumed. Consequently, as Table 2 indicates, there was no decline in per capita food supplies even in Africa during the 1960s and 1970s. Since much of the food imports by Africa during the 1979-81 period was provided by aid, it would have been clearly preferable had the increased food supply come from increased domestic agricultural productivity.

But the most striking evidence of improvement in health and nutrition, even in low income African countries, is the data on increases in life expectancy and declines in infant mortality and child death rates between 1960 and 1982. While the available data are subject to error, the data we do have indicate an increase in life expectancy in low income African countries of 7 to 10 years during the period. This is less than in other low income economies. The increase in life expectancy in India was 12 years and for all low income countries other than China and India the increase was about the same as for Africa, namely 8 years.

The declines in infant mortality and child death rates are clearly important indicators of improvement, though it is obvious that much remains to be accomplished. It is perhaps worth noting that in 1900 the infant mortality rate

Table 1

Estimated Indexes of Food Production Per Capita for Developing Market Economies By Regions, 1966-85
(1961-65 = 100)

	Africa	Latin America	Near East[a]	Far East[a]
1966	96	101	101	94
67	98	104	103	97
68	100	103	103	101
69	100	104	103	103
1970	100	106	103	105
71	99	103	103	103
72	95	101	107	97
73	89	101	99	104
74	93	103	105	99
75	93	103	109	106
76	94	108	111	105
77	87	107	105	106
78	88	110	110	113
79	87	111	107	109
1980	87	112	107	110
81	86	113	108	115
82	86	113	108	112
83	80	110	106	118
84	81	110	106	118
85	86	115	112	120

[1]Source: FAO, *FAO Production Yearbook*, various issues.

[a]The Near East includes Northern Africa and the Middle East.
[b]The Far East includes South, Southeast and East Asia.

Table 2

Per Capita Daily Calorie Supply, World and Regions: 1961-63, 1969-71, 1979-81

Region/Group	Calories per Capita Daily		
	1961-63	1969-71	1979-81
Developed countries	3110	3280	3380
Developed market economies	3080	3260	3370
N. America	3270	3480	3610
W. Europe	3140	3290	3430
Oceania	3190	3280	3150
Other developed	2540	2770	2870
Eastern Europe and U.S.S.R.	3160	3320	3390
Developing countries	2000	2140	2350
Developing market economies	2080	2170	2330
Africa	2130	2180	2260
Latin America	2380	2510	2630
Near East	2290	2410	2840
Far East	1950	2030	2170
Other developing	1950	2190	2310
Asian centrally planned economies	1840	2080	2410
World	2350	2470	2620

Source: Mollett (1985, Table 1, p. 28).

in the United States was 160 per 1000 live births and that in the next 20 years it declined to 80 by 1920. Thus far all low income developing countries both the levels and rate of decline in infant mortality between 1960 and 1980 closely paralleled the U.S. experience in the first two dedades of this century.

The data included in Table 3 support the view that low income countries have generally improved their health and nutrition situation in recent decades. But the data also indicate that there is substantial room for further improvement. It is in this context that approaches that will permit lower cost means of improving food and nutrition merit our support.

Table 3

Life Expectancy at Birth, Infant Mortality Rates and Child Death Rate: Africa, 1960 and 1982

Country Group	Life Expectancy at Birth		Infant Mortality Rates (< 1)		Child Death Rate (ages 1-4)		GNP PER Capita,
	1960	1982	1960	1982	1960	1982	1982 ($)
Africa							
Low income:							
Semi-arid	37	44	203	151	57	34	218
Other	39	49	158	112	37	22	254
Average	38	48	164	117	40	24	249
Middle income:							
Oil importers	41	50	159	111	37	21	670
Oil exporters	39	50	191	113	51	21	889
Sub-Saharan Africa	—	49	170	115	42	23	491
Low-income Economies:							
China	42	67	165	67	26	7	310
India	43	55	165	94	26	11	260
All other	43	51	163	114	—	—	250
Average	42	59	165	87	—	—	280
Middle-income Economics	51	62	126	76	23	10	1,520
Brazil	55	64	118	73	19	8	2,240
Mexico	57	65	91	53	10	4	2,270
Lower-middle income	46	57	144	89	29	13	840
Upper-middle income	56	65	101	58	15	6	2,490
Industrial Market Economies	70	75	29	10	2	—	11,070
Developing Countries by 1982 Per Capita Income:							
Less than $390	42	59	165	87	27	11	280
Excluding China	42	53	—	—	—	—	—
$440-$1,160	50	60	144	89	29	13	840
$1,680-$6,840	56	65	101	58	15	6	2,490

Source: World Bank, *World Development Report 1984*.

The chapters that follow subject the possibilities and prospects of isolated soy protein to a series of exhaustive analyses. It is unlikely that any emerging food product has ever had such careful review in terms of how it fits into prospective developments in food supply and demand, or at least none to my knowledge has had such public display of scholarly discussions. As is readily apparent, the chapters that follow vary in terms of the conclusions reached with respect to the economic viability of isolated soy proteins under varying economic conditions. As is made quite clear, both by those who work for Purina and those who work for academic research institutions, the potential contribution that can be made by isolated soy protein varies depending upon the economic circumstances and the policy objectives of various countries. It was because of this openness of the discussion and the knowledge that whatever the results might be, the results would be published that induced me to have a modest role in the evaluation.

It is not my intention in this brief chapter to review the various contributions to this volume. Based upon these papers, I have been asked to review some of the issues and conclusions in a policy context. The primary policy issues are these: First, are isolated soy proteins lower cost in providing nutrition than meat and other animal products? Second, are the products that utilize isolated soy protein capable of obtaining ready acceptance by consumers when there is full freedom of selection? Third, how do the particular economic circumstances of a country and its economic policy objectives influence the economic viability of soy proteins?

These three issues are addressed in the chapters that follow. And it seems to me that they have been addressed in an appropriate manner. Chapter 6 addresses the first question, not so much in terms of providing an answer that has general applicability but by illustrating the appropriate methodology for determining if soy proteins are a potential way of increasing the supply of meat products at lower cost than by increasing the production of meat products. The particular results that are presented are a function of the technical and economic assumptions. The author correctly states: "It must be realized that what follows is only an illustration of how to carry out the analysis. It is not meant to imply that what is shown here as most appropriate is necessarily so in all cases. The ranking of the profitability of the alternative depends upon the technical coefficients, prices and government interventions. Each of these areas varies considerably from country to country and from case to case within countries." The chapter presents a clear and relatively simple approach for comparing the alternatives.

It is novel for an exercise of this kind to show how various distortions resulting from governmental policies may affect the results. Most governments have a variety of taxes and subsidies that are designed to influence the financial viability of an activity within the country. However, these taxes and subsidies confuse the issue of the social or true profitabilities of alternative processes for achieving the same objective. Even when there are relatively modest distortions, as in the examples given, the existence of taxes and subsidies affect the relatively profitability of different activities from the standpoint of the economy compared to the profitability of a private enterprise that is affected by the taxes and subsidies.The analysis is particularly noteworthy in including the effect of distortions introduced by an overvalued exchange rate. An overvalued currency acts like a subsidy for imports and a tax on exports and thus affects financial comparisons of alternatives that have different mixes of imported inputs or products that may be exported. Comparing the cost estimates given inTable 4 and Table 8 show the effects on true economic costs of a rather modest overvaluation of the foreign exchange rate.

Chapter 7 summarizes scores of studies measuring the relative values of various protein food sources. Studies of major scientific institutions are included as well as studies published by researchers in refereed scientific journals. These studies establish that soybean proteins have many desirable characteristics and few limitations in their role as a supplement to meat products. The chapter notes that there are a number of sources of vegetable proteins but soybeans are one of the lower cost sources and has the advantage of already developed technologies for processing. These facts are important in evaluating the viability of isolate soybean protein.

Chapter 8 presents a very interesting analysis of why some new foods or variations of existing food succeed or fail to find ready acceptance by consumers. The authors consider several possible approaches to introducing vegetable proteins—in meats, in baked products and as beverages. There have been successes and failures in each. What seems clear is that if appropriate recognition is given to the various attributes of food products that people evaluate in making their decisions, the probability of acceptance of a product contain isolate soybean protein or similar protein product can be increased to a high level.

Chapter 2 presents a set of projections of meat supply and demand in developing countries to the year 2000. The projections imply an enormous increase in the potential gap between meat consumption and production by the end of the current century. The projections included here are at variance to those made by the Food and Agricultural Organization in Agriculture: Toward 2000 (AT 2000). In AT 2000 three projections are made of the net trade balance for meat of 90 developing countries. For a projection based on trends from 1961-65 to 1980, the developing countries would have a meat trade export balance or 3 million metric tons. Two other projections are based on detailed analyses of production prospects and demand possibilities reflecting income and population projections show approximate balance in meat trade for the developing countries. The projections in Chapter 2 indicate that desired consumption would exceed trend production by nearly 21 million tons. For comparison, total world trade in meat in 1981 was a little more than 8 million tons.

What is at least as relevant as projections of meat production and demand is what may be expected to be the price trends for meat relative to soybean and cereals. Recent trends have been diverse. Poultry meat prices have fallen in recent years relative to grain and soybean prices while beef and pork prices have increased relative to the grains and soybeans. If the long run costs of producing bovine and ovine meat increase relative to cereals and soybeans, this should have a positive effect upon the demand for isolated soybean protein. Thus policy analysis should consider prospective price trends of competing products. The price trends could be favorable to the growth of use of soy proteins even if the FAO projections of the trade balance for meats in the developing countries turns out to be reasonably accurate. While neither the FAO nor the IFPRI projections include any adjustment for possible price changes, the increase in total meat production in the developing countries by 2000 could be large enough to result in higher relative costs of production than prevailed in the late 1970s.

Chapters 3, 4 and 5 include interesting case studies of the potential demands for isolated soy protein. The countries differ substantially from each other— Sweden is a high income industrial country, Mexico is a middle income developing country and China is the world's most populous country and one of the poorest ones. In recent years, China has undergone major policy reforms in the agricultural sector and has shifted from being a major importer of grains for nearly two decades to a net exporter in 1985/86.

These studies illustrate a number of the important issues raised in Chapter 6. It appears that the profitability of soy protein in Sweden represents financial and not economic or social profitability. Due to a combination of domestic agricultural and trade policies, meat prices to both consumers and producers are substantially in excess of international market prices and especially so for beef. Thus soy protein may be financially profitable without resulting in a saving of resources for the nation.

The very interesting case study for China indicates that under the conditions that prevailed during the first half of the 1980s the importation of isolated soy protein would have been economical for China. This conclusion rested upon the relative costs of importing corn and the soy protein, the feedgrain:hog conversion ratio and the potential savings of domestic transport and marketing resources through the use of soy protein. By the mid-1980s China had become a modest net exporter of grain, including corn. Thus the corn price that should be used became the realized export price of China's corn and this price is substantially lower—perhaps by a quarter or more—than the import cost. In addition to the shift in China's trade position for corn, the international market price of corn has declined substantially though surely there will be recovery from the low prices prevailing in late 1986.

While the Mexican case study is less detailed than the other two, it indicates several reasons why soy protein may have either financial or economic feasibility. In Mexico the existence of a large number of low income families raises the possibility of undernutrition and emphasizes the desirability of measures that will reduce the cost of an adequate diet. Another feature of the

Mexican economy is the limited capacity of its transportation system; importing large quantities of feed grains imposes substantial burdens upon that system. Given the very large external debt that Mexico has and the necessity that it faces of achieving a very large current account balance, the current Mexican situation may represent one in which it is appropriate to give greater weight to savings of foreign exchange than is implied by even an equilibrium exchange rate. This would be the case, for example, if more rapid pay down of principal resulted in a reduction of future interest rates charged on that debt.

The available evidence strongly supports the conclusion that isolated soy protein has the potential for providing an increase of available food supplies by supplying a food product with high nutrition values while lowering the overall cost of meat products. The soy protein may have particular value in improving the nutrition of low income groups. This conclusion is only hinted at in the existing studies and seems worthy of further exploration.

If soy proteins are to play the role in a nation's food supply that is economically appropriate, legal and institutional barriers to their utilization should be kept to a minimum. Soy proteins should have the opportunity to win their place in national diets based on their nutritional contribution and acceptance and relative costs. This means that barriers to investment in production facilities should be minimized and the end products should not be subjected in differential rates of taxation. In many countries the available market for soy proteins will not support national production of the product. This will mean that at least for some years it will be necessary to import the soy proteins though the mixing with meat and other products would occur in the country where the consumption occurs. In these cases, the importation of soy proteins should not be subjected to greater barriers than affect the importation of feed grains and other feeding materials.

Chapter 2

Meat Supply and Demand in Developing Countries: Past Trends and Projections to 2000

*J. S. Sarma**

*International Food Policy
Research Institute
Washington, DC*

I wish to acknowledge gratefully the advice and assistance given by Leonardo Paulino, the Program Director, in the preparation of the paper. We had also useful discussions with F.H. Schwarz and James K. Allwood of the Ralston Purina Company with regard to the scope and content of the study. I also thank Darunee Kunchai for her competent assistance in the tabulation and analysis of data and Frances Walther for typing of the manuscripts.

*J.S. Sarma is Research Fellow, IFPRI.

11

Abstract

Animal products in various forms of meat, eggs and milk products supply about one-sixth of the calories and one-third of the protein in per capita human food supplies in the world. When income increases, people change their diets to increase consumption of animal products.

High consumption of animal products is established in developed countries. With rising per capita incomes, however, demand for these commodities in Third World countries will rise faster than in developed countries. This trend has major implications for livestock and grain production.

The analysis presented in this chapter covers 104 developing countries. Twenty-one are in Asia, 19 in North Africa/Middle East, 40 in Sub-Saharan Africa and 24 in Latin America. The author presents an overview of the supply-demand balances for meat and projections to 1990 and 2000.

It is clear, using either strong or weak economic growth assumptions, that there will be major shortfalls in supplies of animal products relative to demand in these countries.

The output of meat from the early 1960's (1961-65) to the mid-1970's (1973-77) increased in these countries at an average annual rate of 2.9 percent, faster than the corresponding population increase of 2.6 percent. Rising per capita income, however, stimulated meat consumption to rise faster, at 3.2 per cent annually.

Net meat exports from these countries consequently declined from 760,000 tons to 304,000 tons over the same period. Developing countries became net importers of meat by the end of the 1970's. If past economic growth rates continue, the aggregate projected demand for meat is expected to outstrip projected supply by about 8 million tons by 1990 and by 21 million tons by 2000.

World production of meat in 1981 totaled 140 million tons. The 104 countries studied here, which account for half of the world's population, produced less than one-fifth of the total meat output.

Assuming continuation of historical trends, Third World countries are expected to produce 36 million tons of meat by 1990. Fifty percent of the increase over the 1977 level would be in Latin America and about 20 percent in Asia.

The total requirement of meat in these countries in 1990, however, is projected by 44 million tons, 8 million tons more than projected production. Forty-three percent of the projected increase of 20.6 million tons in consumption between 1977 and 1990 would be from population increases, with the balance attributable to increasing per capita income.

If the projected demand for meat in the developing countries can be met, it could total 72 million tons, or an average annual per capita consumption of 20 kilograms by the year 2000. Meat production in Third World countries, however, is projected at 51 millions tons in 2000, leaving a gap of 21 million tons.

Unless this gap is bridged through accelerated production growth or large-scale transfer from the developed countries, the relative prices of meat in the Third World will rise. This would cause considerable hardship to poor consumers, especially in countries where meat constitutes a significant portion of food intake.

Of the projected supply-demand gap of 21 million tons of meat by 2000, North Africa/Middle East and Asia together would share nearly two-thirds. The projected deficit in Latin America would be relatively small—2.7 million tons; Sub-Saharan Africa would have a shortfall of 4.7 million tons. East and Southeast Asia and Western Asia would have large gaps of about 5 million tons each, and deficits in South Asia and Northern Africa would be about 1.8 million tons each. The gap in Upper South America would be 3.6 million tons.

These are the areas where meat supplies need to be augmented. The projected demand has major implications for government policy, grain production, animal husbandry, food processing technology, and international trade.

Parallel large supply-demand gaps could occur in feed grains if past trends in production and per capita income also continue to 2000.

Policies and technologies for achieving the needed increases in meat supply are examined by other chapters in this volume.

Introduction

With economic growth, livestock production in developing countries assumes major importance in food supply-demand relationships. The income elasticity of demand for animal products in these countries is high not only relative to that of cereals, but also as compared to the corresponding elasticity in the developed economies. This implies that with rising per capita incomes, the demand for these commodities would rise faster in the Third World countries and in turn would influence the derived demand for cereals and other staple foods as livestock feed. Even at present, animal products contribute about one-sixth of the calories and one-third of the proteins in the per capita food supplies in the world. Calorie and protein levels from livestock products in the developing countries are much lower than those in the developed countries.

Results of IFPRI's food gap studies suggest that the rapid growth of incomes in many developing countries would greatly accelerate food consumption growth in the Third World during coming decades. Increases in the use of the basic food staples for direct human consumption due to growth of population and per capita incomes will be augmented by the fast-rising derived demand for these commodities for feed arising from the changes in the diet patterns toward more animal products, especially meat. This can be expected to lead to significant increases in basic staple imports, particularly cereals, and/or shifts of resources to feedgrain production in the Third World countries. An analysis of the supply-demand relationships specifically of meat in developing countries that would identify areas where serious problems of imbalance are likely to arise can help guide plans designed to ease these problems.

Background of study

The role of isolated soy protein as an ingredient in meat products to increase national food supplies is the subject of research sponsored by the Ralston Purina Company. Questions are raised on the nature and magnitude of the benefits of such form of soy protein utilization and whether this or other uses of soy protein products would be of value to middle and lower-income people and countries. Ralston Purina Company requested the International Food Policy Research Institute (IFPRI) to look into the supply-demand relationships of meat in Third World countries. Results of the analysis are expected to provide answers to some of the issues relating to the future role of isolated soy protein as an ingredient in meat products.

Objective

The general objective of this study is to analyze the trends in the production, consumption and trade of meat in the Third World. Specific objectives include the examination of the past and current levels of meat supply and demand imbalances in developing countries, and presenting a trend-based scenario on the levels of these imbalances by 1990 and, more generally, by 2000. In addition, an alternative scenario with per capita income growth 25 percent less than the trend growth is also presented. The study also looks into the

15

trends in the utilization of cereals as animal feed in Third World countries, in the context of the overall situation regarding the supply-demand balances for cereal grains.

Data and Methodology

The study avails largely of the Food and Agriculture Organization data base used for IFPRI's work on trends in livestock and poultry products and in the basic food staples of developing countries. Data on supply-demand balances are obtained for 1961-65 and 1973-77[1] and their projections to 1990 and 2000 (in broad aggregates) are derived from those of production and consumption. The former are worked out as trend extrapolations of the period 1961 to 1977 while the projections of consumption are obtained taking into account trend estimates of per capita consumption, 1977, income elasticity of demand and trend growth/25 percent less than trend growth of per capita income between 1966 and 1977. A note on the data and methodology adopted for the study is given in Appendix C.

Although the trends and projections analysis is done by country, study results are presented for country groups based on geographical regions and sub-regions and GNP/capita levels. The analysis in this report covers 104 developing countries of which 21 are in Asia, 19 in North Africa/Middle East, 40 in Sub-Saharan Africa and 24 in Latin America.[2] A list of developing countries included in each region/sub-region and per capita income group is given in Appendix B.

An overview of the supply-demand balances for meat and the projections to 1990 and 2000, classified by different regions, is given in Section 2. The highlights of the past trends and future projections of production, consumption and trade/balance by groups of countries are discussed in Sections 3, 4 and 5, respectively. The production and consumption of meat by type of meat are analyzed in Section 6. Section 7 deals with past trends and projections to 1990 and 2000 of cereal feed. A brief but similar analysis of supply-demand relationships for grains is given in Section 8. Section 9 attempts to indicate briefly the policy implications of the meat and cereal supply utilization trends.

[1]The latest available data on consumption of meat relate to 1977.

[2]The study excludes China from its trends analysis because of data problems. However, for presenting the global picture, FAO data on China are utilized for consistency and completeness.

Supply-Demand Balances - Meat

The analysis of past trends of production, consumption and trade in meat recently completed in IFPRI indicates that for the 104 developing countries covered by the study, the output of meat[3] increased between the early sixties (1961-65) and the mid-seventies (1973-77) at an average annual rate of 2.9 percent, which is higher than the corresponding population growth rate of 2.6 percent. Yet, because of the effect of rising per capita incomes, the consumption of meat rose faster at 3.2 percent per annum. Consequently, the net meat exports from these countries declined from 760 thousand tons[4] to 304 thousand tons over the same period. Later figures of trade in meat show that developing countries have become net importers by the end of the 1970s. If the past rate of economic growth continues into the future, even if the past output growth is maintained, the aggregate projected demand for meat is expected to outstrip the projected supply by around 8 million tons by 1990 and by 21 million tons by 2000.[5]

In 1981, the total world production of meat amounted to 140 million tons comprising 47 million tons of beef and buffalo meat, 8 million tons of mutton and goat meat, 56 million tons of pig meat and 29 million tons of poultry meat. The 104 study countries which accounted for half of the world population produced a little less than one fifth of the total meat output. Within the developing world, there were wide differences in the total and per capita levels of output among the different regions (See Figure 1).

There were 1.33 billion cattle and buffaloes, 1.6 billion sheep and goats, 777 million pigs and 6.74 billion chickens, ducks and turkeys in the world in 1981. Nearly two-thirds of the cattle and buffaloes, a little more than half of the sheep and goats, one sixth of the pigs and about two-fifths of the poultry were in the study countries (See Figure 2).

In the mid-seventies nearly 20 percent of the cattle population[6] were slaughtered in the world for meat every year, varying from one-third in the developed to one-tenth in the developing countries. A higher proportion of sheep and goats are slaughtered every year (35 percent) although the differences in the slaughter percentages for sheep are narrower than those for goats between the developed and the developing economies. The slaughter percentage of cattle is the highest in North Africa/Middle East, followed by Latin America and the lowest in Asia (See Table 1).

Average carcass weights of beef and veal and pork per slaughtered animal in the study countries are about 70 to 75 percent of the levels in the developed countries. The differences in the case of mutton and goat meat between the two groups of countries are much narrower (See Table 2).

---

[3]Comprising beef and buffalo meat, mutton and goat meat, pig meat and poultry meat expressed in carcass weight equivalents.

[4]Tons refer to metric tons in this chapter.

[5]See "Livestock Products in the Third World - Past Trends and Projections to 1990 and 2000" -International Food Policy Research Institute, Washington, D.C., April, 1985.

[6]The cattle population refers to the total stock including bulls, bullocks, cows and calves.

Figure 1
World Population:1981
(percentages)

World Meat Production
1981 : 140 Million Metric Tons

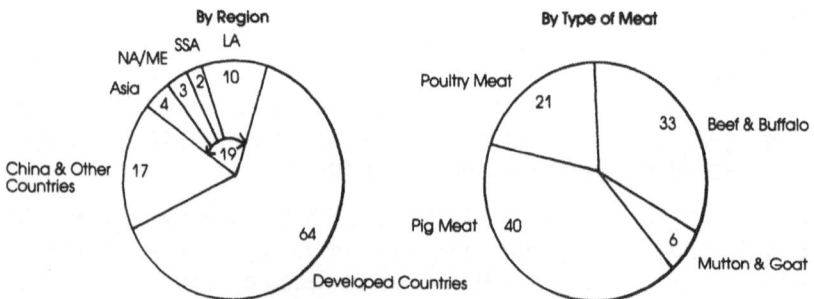

Comparison of World Meat Output
Per Capita:By Region:1981 (kilogram)

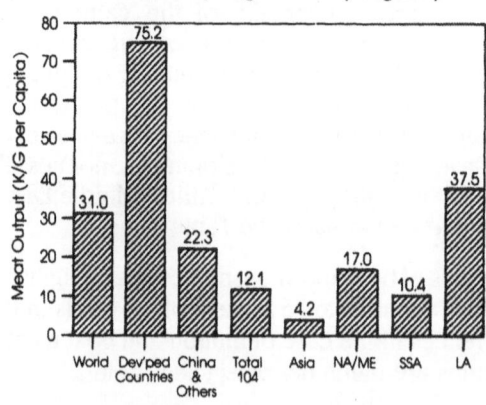

NA/ME = North Africa/Middle East   SSA = Sub-Saharan Africa   LA = Latin America

Sources: Food and Agriculture Organization of the United Nations, FAO Production
Yearbook, 1982, Rome 1983.
United Nations Department of International Economic and Social Affairs, World
Population Trends and Prospects by Country 1950-2000 (ST/ESA/SER.R/33, 1979)

Figure 2
World Livestock Numbers: By Region
1981

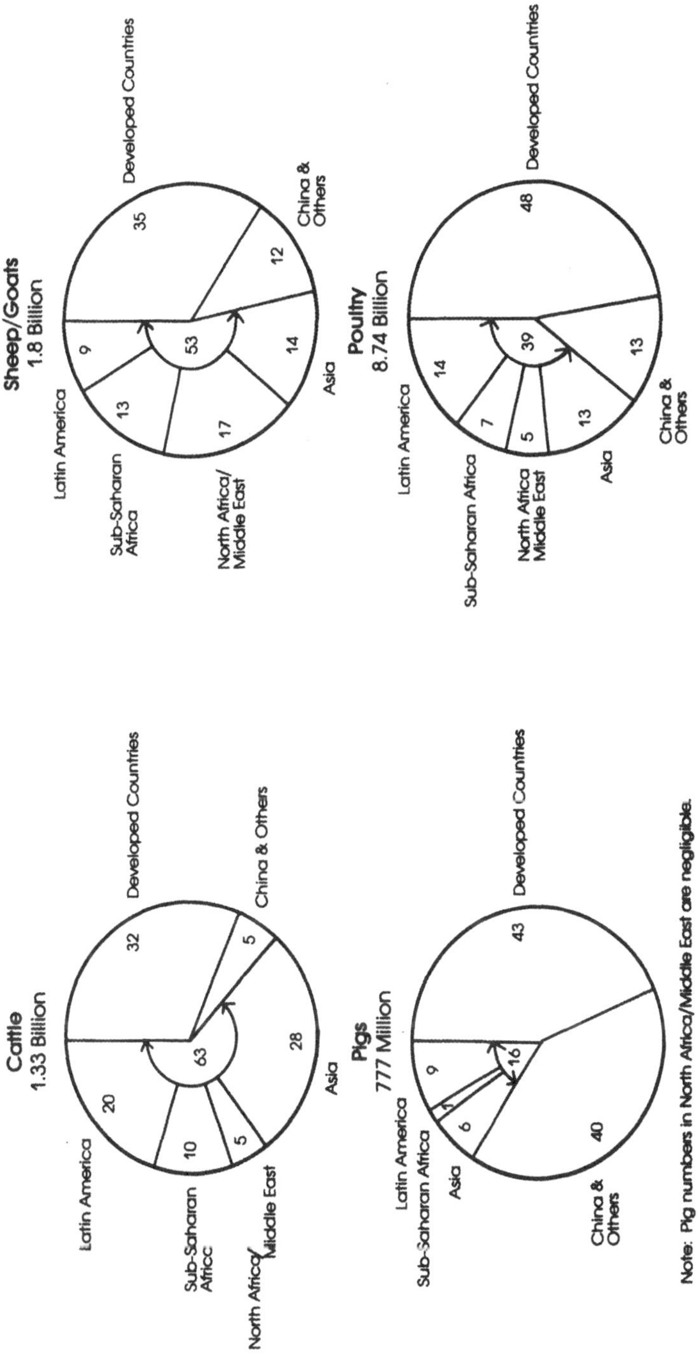

Cattle
1.33 Billion

Latin America 20
Sub-Saharan Africa 10
North Africa/Middle East 5
Asia 28
China & Others 5
Developed Countries 32
63

Sheep/Goats
1.8 Billion

Latin America 9
Sub-Saharan Africa 13
North Africa/Middle East 17
Asia 14
China & Others 12
Developed Countries 35
53

Pigs
777 Million

Latin America 9
Sub-Saharan Africa 1
Asia 6
China & Others 40
Developed Countries 43
16

Poultry
8.74 Billion

Latin America 14
Sub-Saharan Africa 7
North Africa Middle East 5
Asia 13
China & Others 13
Developed Countries 48
39

Note: Pig numbers in North Africa/Middle East are negligible.

Source: Food and Agriculture Organization of the United Nations, FAO Production Yearbook, 1982, Rome 1983.

Table 1 Percentage of animals slaughtered[a] for meat, average 1973–77

Country Group	Cattle	Buffalo	Sheep	Goat	Pigs[b]
World[c]	18.9	5.7	34.9	34.7	95.5
Developed Economies	33.6	9.9	40.6	51.5	126.1
Developing Economies	10.3	5.7	29.1	33.6	71.0
104 Study Countries[d]	9.7	5.3	29.1	35.5	56.8
Asia	3.3	4.5	33.9	43.2	73.7
North Africa/ Middle East	15.8	27.6	34.4	32.3	99.3
Sub-Saharan Africa	10.1	*	26.9	32.1	74.9
Latin America	14.7	*	19.0	26.2	43.5

[a] Slaughter percentage is obtained by dividing the number of animals slaughtered (which include the net imports of live animals into the country) by the total of indigenous stock plus net imports or minus net exports of live animals, of each species and multiplying by one hundred.

[b] The percentage of pigs slaughtered exceeds 100 in some countries because more than one crop of pigs is raised and slaughtered in a year.

[c] Country grouping follows FAO classification.

[d] Country grouping follows IFPRI classification.

* Negligible.

Source: *Food and Agriculture Organization of the United Nations, FAO Production Yearbook Tape, 1979, Rome, 1980.*

Table 2 Average output of meat per animal, average 1973–77 (in kilograms)

	Beef & Veal	Buffalo Meat	Mutton & Lamb	Goat Meat	Pig Meat
World[a]	194.1	155.4	15.2	11.4	68.0
Developed	212.2	188.2	16.0	11.7	76.6
Developing	159.5	155.0	14.1	11.4	55.8
104 Study Countries[b]	162.3	130.2	14.0	10.8	54.8
Asia	125.8	104.0	11.8	9.9	47.0
North Africa/ Middle East	112.6	133.2	15.9	14.1	66.9
Sub-Saharan Africa	125.0	—	11.0	9.8	45.2
Latin America	193.6	—	14.5	11.0	65.0

[a] Country grouping follows FAO classification.
[b] Country grouping follows IFPRI classification.

Source: Food and Agriculture Organization of the United Nations, FAO Production Yearbook Tape, 1979, Rome, 1980.

Trends in meat output

The output of meat in the study countries increased from 15.6 million tons in the early sixties to 22 million tons in the mid seventies or by nearly 40 percent over this period. The distribution of the output increases, has been, however, rather uneven among the different regions. In Sub-Saharan Africa, for instance, meat production lagged behind population growth, although in the other three regions the growth in the former exceeded that of population.

Figure 3
Total Meat Production: By Region
1961-65 - 1973-77 (Million Metric Tons)

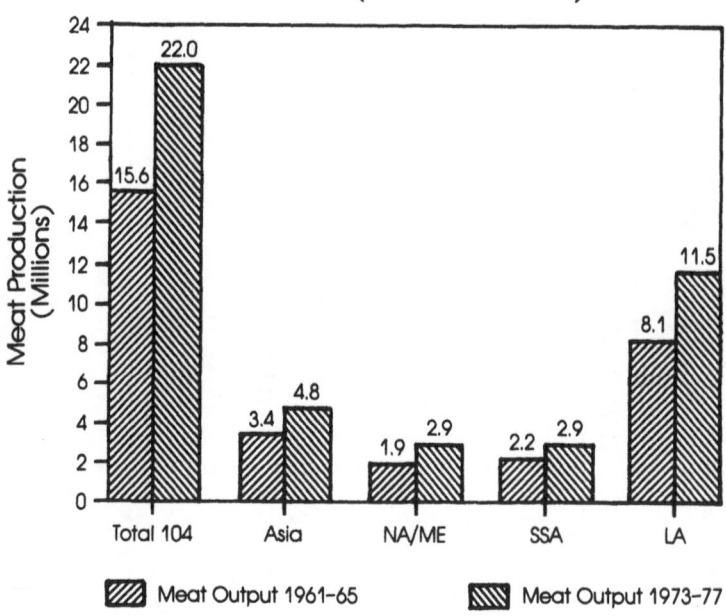

Source: Food and Agriculture Organization of the United Nations, "Global Agricultural Programming System Supply Utilization Accounts Tape," Rome, June 1980.

In the early sixties, total ruminant production constituted nearly three fourths of the total meat output in the developing countries. By the mid-seventies, the combined share of poultry and pigmeat increased from 26 to 34 percent at the expense of ruminant meat, because of the relative ease with which the former meats could be increased through the adoption of better breeding and feeding practices. In particular, the share of poultry in total meat production doubled in Latin America and North Africa/Middle East while in Sub-Saharan Africa its proportion increased to nearly one and a half times.

22

Trends in meat consumption

During the mid-seventies the study countries consumed around 21.7 million tons of meat as compared to 14.9 million tons in the early sixties. Asia with 58 percent of the population consumed 23 percent of the meat while Latin America, with 17 percent of the population, consumed 50 percent. Sub-Saharan Africa consumed the least amounts of meat.

Figure 4

Total Meat Consumption: By Region

1961–65 - 1973–77 (Million Metric Tons)

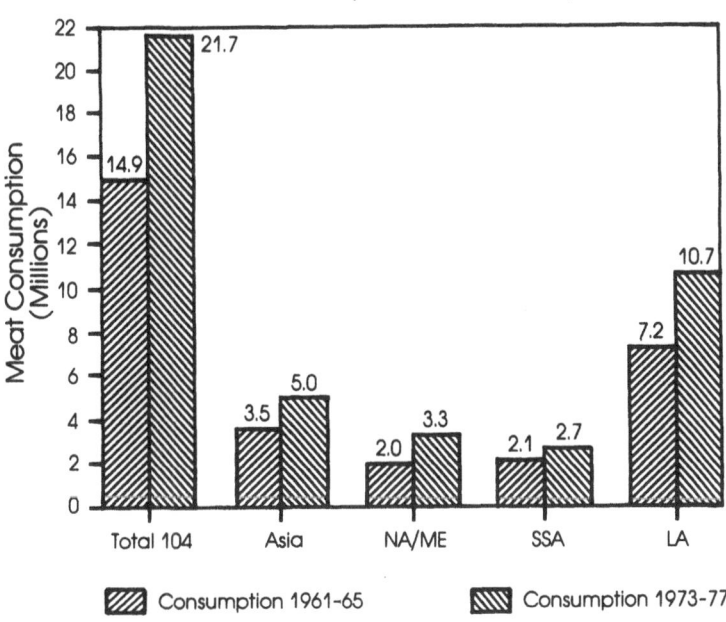

Source: Food and Agriculture Organization of the United Nations, "Global Agricultural Programming System Supply Utilization Accounts Tape," Rome, June 1980.

Among the regions, North Africa/Middle East exhibited the most rapid growth while Sub-Saharan Africa had the slowest consumption growth rate.

The composition of meat consumption followed roughly the regional pattern of production. In the early sixties ruminant meat formed about 72 percent of the total meat and the balance of 28 percent represented poultry meat and pork. By the mid-seventies the share of the former declined to 64 percent of the total. Poultry and pig meats increased their share to 18 percent each. This showed a decrease of 8 percentage points in the consumption of large and small ruminant meats over this period.

Relationship between meat consumption and levels of income

An analysis of the relationship between the levels of per capita income and per capita consumption of livestock products indicated that for all the study countries taken together, meat consumption ranged from 4 kilograms in the very low income countries (less than $250 per capita) to 35 kilograms per annum in the high income group with per capita income greater than $1,250.

Table 3. Per capita consumption of meat, average 1973-77 by level of per capita income in 1977

(kg/year)

Per Capita Income Level	Asia	North Africa/ Middle East	Sub-Saharan Africa	Latin America	104 Study Countries
Less than $250	2.8	8.9	10.2	10.6	4.0
$250-$499	6.9	12.4	7.5	19.2	8.1
$500 - $1,249	10.2	14.6	14.1	22.8	15.9
Greater than $1,250	56.8	18.3	27.4	38.5	35.0
All groups	4.3	14.4	9.3	33.8	10.9

Source: Food and Agriculture Organization of the United Nations, "Global Agricultural Programming System Supply Utilization Accounts Tape," Rome, June 1980.

Trends in Meat Trade

Between the early sixties and the mid-seventies, meat exports of the study countries increased marginally from 1.5 to 1.6 million tons, while the meat imports rose almost by 80 percent from 0.7 to 1.3 million tons. Consequently, the net trade surplus of the Third World countries declined by nearly 456 thousand tons to around 304 thousand tons during this period. In the mid-seventies, Latin America had a net surplus of 0.9 million tons while both Asia and North Africa/Middle East had an average annual deficit of 0.3 million tons each. In both periods, 70 percent of the meat exports from Third World countries were from Latin America. Imports of meat into North Africa/Middle East rose from 0.14 million tons to 0.40 million tons, while those into Asia nearly doubled over this period.

Figure 5
Trends in Trade in Meat

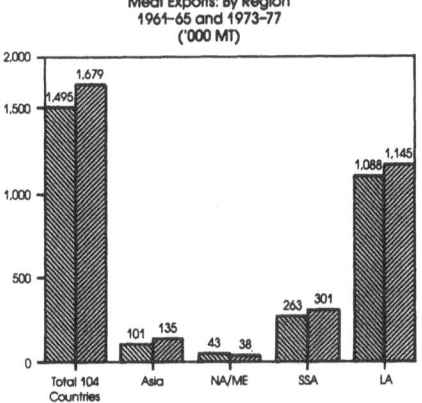

Meat Exports: By Region
1961–65 and 1973–77
('000 MT)

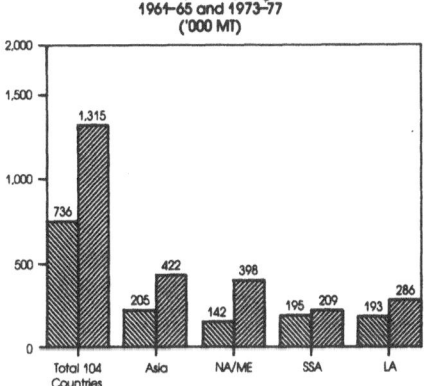

Meat Imports: By Region
1961–65 and 1973–77
('000 MT)

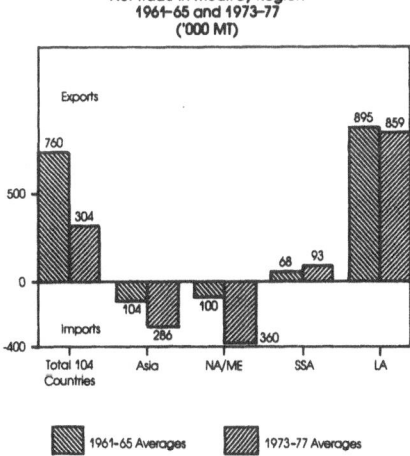

Net Trade in Meat: By Region
1961–65 and 1973–77
('000 MT)

1961-65 Averages   1973-77 Averages

Source: Food and Agriculture Organization of the United Nations, "Global Agricultural
Programming System Supply Utilization Accounts Tape," Rome, June 1980.

If these trends continued, meat surpluses within the developing countries would not be adequate to meet the demand from the meat deficit countries within the region in the coming years.

Projections for 1990 and 2000

If historical trends in the output of meat continue, Third World countries are projected to produce 36 million tons of meat by 1990. Fifty percent of the increase over that in 1977 would be in Latin America and about 20 percent in Asia, the balance being shared by North Africa/Middle East and Sub-Saharan Africa.

Based on the assumptions regarding growth of population and per capita income, and the available estimates of income elasticity of demand, which are spelled out in Appendix D, the total requirement of meat in 1990 is estimated to be around 44 million tons. Thus the gap between the projected production and consumption in that year would be 8 million tons. Forty-three percent of the projected increase of 20.6 million tons in consumption between 1977 and 1990 would be from growth in population and the balance due to increasing per capita incomes. The largest gap would be in Sub-Saharan Africa (3.0 million tons) and the smallest in Latin America (0.9 million tons). Projected consumption in Asia and Sub-Saharan Africa would fall short of the respective outputs by 2.7 and 1.4 million tons, respectively.

Figure 6

Projections of Meat: 1990

Production/Consumption (Million Metric Tons)

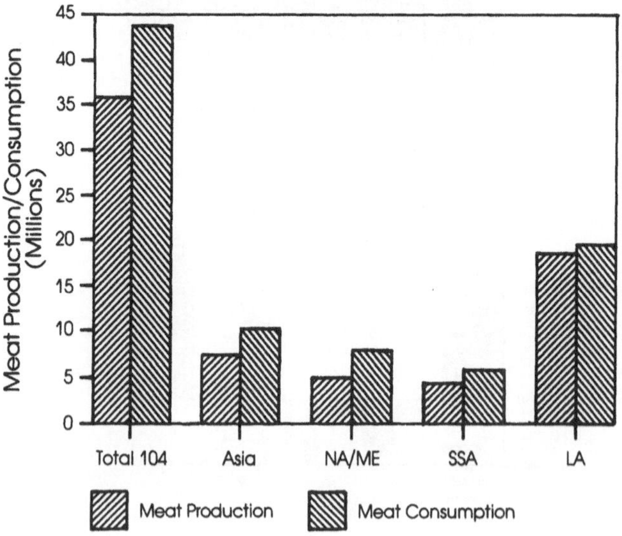

Source: IFPRI projections.

If the projected demand for meat could be met either through increased production within the developing countries or through imports from developed countries, the average annual per capita consumption would reach 20 kilograms for the 104 study countries by 2000. The per capita consumption in Latin America would be the highest at 49 kg. while the lowest would be in Asia at 9 kg.

Table 4. Projected levels of per capita human consumption of meat, 1990 and 2000

Region	1990	2000	(kg/year)
Asia	6.3	8.7	
North Africa/Middle East	23.3	32.5	
Sub-Saharan Africa	13.2	17.9	
Latin America	41.0	48.8	
104 Study Countries	15.1	19.7	

Source: IFPRI Projections.

By 2000, if these trends continue, the Third World countries could produce 51 million tons of meat which would be around 2.2 times the 1977 trend level; while the demand for meat is projected to rise to 72 million tons, the gap between the projected production and consumption would widen to 21 million tons. A regional breakdown of these figures indicates that even the

Figure 7

Projections of Meat: 2000

Production/Consumption (Million Metric Tons)

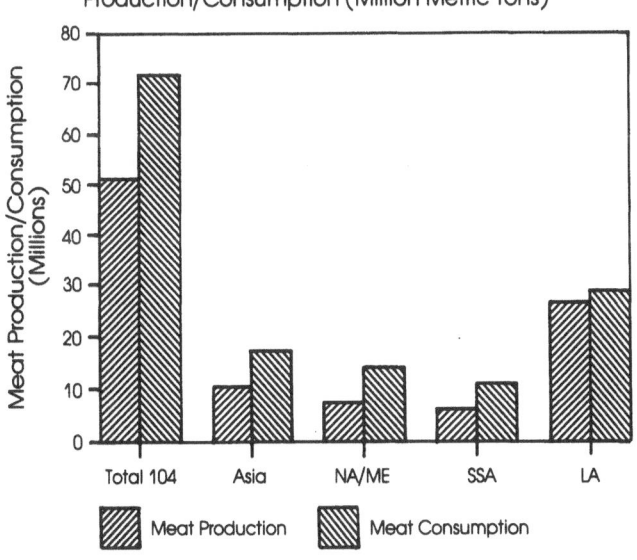

Source: IFPRI projections.

27

trend-based projections of out-put would nowhere meet the anticipated demand under constant price assumptions either at the regional level or for the 104 countries taken together.

Unless this gap is bridged through accelerated growth in production or large-scale transfers from the developed to developing countries through international trade, the relative prices of meat would rise which in some cases would cause considerable hardship to the poor consumers, particularly in countries where meat forms a considerable proportion of the food intake. As large-scale imports into developing countries might not be feasible for countries with foreign exchange and other constraints, redoubled efforts are needed to augment the output of meat.

Estimates of Consumption Based on Low Income Growth

The estimates of consumption and of surpluses and deficits discussed in this section so far are based on the assumption of a continuation of 1966-77 trends in income which were high in several countries due to a variety of factors discussed in the note on "The Population and Income Growth Assumption in the Projections" in the Appendix D. Subsequently, however, economic growth in many of these countries has slowed down because of the two oil shocks of 1974 and 1979-80, the ensuing recession and protectionism in the OECD countries, and mounting debt repayment problems. However, there are several other developments which enhance prospects of better performance in the future. Keeping this in view alternative projections of meat consumption on the basis of slower income growth have been worked out for 1990 and 2000 and are given in Table 2.3, Appendix A. These projections indicate that the total consumption of meat would rise to 39.7 million tons in 1990 and 61.6 million tons in 2000. If the past trends in meat output continue, the Third World countries are projected to have deficits of the order of 4 and 10 million tons of meat in 1990 and 2000 respectively. Latin America would be surplus in meat while the other regions would continue to be in deficit.

Meat Output: Trends and Projections

Latin America is the largest meat producing region in the Third World accounting for a little over fifty percent of its output. Projections based on country trend growth rates indicate that Latin America will maintain its dominance in 1990 and 2000 also. The most rapid growth between the early sixties and mid-seventies was, however, in North Africa/Middle East, where meat output rose at 3.5 percent per annum. This rate is projected to be accelerated further to 4 percent between 1977 and 2000. (Table 1.1, Appendix A)

Among the sub-regions, the annual increase in meat output was the most rapid in Central Africa at 4.8 percent, which was more than double that for Sub-Saharan Africa. The projected output in this sub-region in 2000 would be three times that in 1977, although even at this level, it would be only one-seventh of that for the region as a whole. Around thirty percent of meat in the developing countries is projected to be produced in Upper South America alone in 2000, an increase of 5 percentage points from its 1977 share. Lower South America, because of its slower growth rate (1.4 percent per annum) would show a decline from 17 percent to 11 percent over the same period. Similarly, because of differences in the projected growth rates, the meat output in East and Southeast Asia which was double that in South Asia in 1977 would be three times that of the latter in 2000.

Meat output based on a classification of countries by per capita income levels revealed wide differences in the per capita output in the different income groups. For example, in the mid-seventies, nearly half of the meat in the Third World was produced in countries with income levels of $1,250 and above, which accounted for about 15 percent of the Third World population, whereas the very low income countries (less than $250) with more than half the population shared less than a fifth of the output. The annual increase of meat output in the latter group was also low (1.9 percent) because of which their share in the projected output in 2000 would decline to 15 percent. The high income countries of Latin America produced nearly 92 percent of meat output of all the developing countries in this group in the mid-seventies. The most rapid increase was, however, achieved by the high-income countries in North Africa/Middle East (5.7 percent) over the reference period. (Table 1.2, Appendix A)

Meat consumption: Trends and Projections

The regional and sub-regional distribution of meat consumption generally followed the pattern of production discussed in section 3 as modified by trade to be discussed in the next section. In the mid-seventies, nearly half of the Third World meat consumption was in Latin America. By 2000, however, the share of Latin America in total meat consumption is projected to decline to 40 percent although per capita consumption of the region would still be higher than in the mid-seventies. Consumption growth was most rapid in North Africa/ Middle East, because of rapid increase in income associated with oil-revenues in several countries of the region, and is projected to accelerate further during the period 1977-2000 if the past income growth among countries continues. The share of this region in meat consumption which was one-seventh in 1977 is projected to increase to one-fifth of the Third World total by 2000. (Table 2.1, Appendix A)

Among the sub-regions, Western Asia exhibited the most rapid growth in meat consumption and South Asia the slowest. The consumption growth in South Asia was influenced by the relatively large weight of India which had a slow increase in meat consumption between the early 1960s and the mid-1970s. In Sub-Saharan Africa, past trends in meat consumption indicate a slow growth of about 2 percent per annum as compared to a population growth of 3 percent per annum implying declining per capita meat consumption, particularly in West Africa. If the 1966-77 trends in per capita income continue, the meat consumption in West Africa would increase at more than 8 percent per annum, mainly because of the rapid growth expected in Nigeria.

As observed in the case of output, the high income countries had a share of 50 percent of the total meat consumption in the Third World in the mid-seventies. The rate of increase was particularly rapid in the high income countries of North Africa/Middle East (8.3 percent), Sub-Saharan Africa (6.3 percent) and Asia (5.7 percent) between the early sixties and mid-seventies. But the projected growth rates for 1977 to 2000 are lower except in Latin America. However, because of the large weight of this region in the group of high-income countries, the overall growth rate for this group would still be 4.3 percent over the period 1977 to 2000. (Table 2.2, Appendix A)

At the other end, low income countries as a group shared less than one-fifth of total meat consumption in the Third World in the mid-seventies and their rate of increase also was slow - (less than 2 percent per annum). If the past trends in per capita incomes continue, the annual increase in meat consumption of the low income countries would double to nearly 4 percent per annum but, because of faster increases in other regions, their share in the total meat consumption would go down to 14 percent in the year 2000.

Meat Trade - Trends and Projected Supply-Demand Balances

Latin America was a major net exporter of meat in the mid-seventies (859 thousand tons) followed by Sub-Saharan Africa (92 thousand tons). Asia and North Africa/Middle East were net importers (286 thousand tons and 360 thousand tons respectively). Between the early sixties and the mid-seventies, aggregate meat exports from Latin America increased from 1.09 million tons to 1.14 million tons. Within the region, Upper South America increased its exports by nearly two hundred thousand tons and Central America and Caribbean slightly by 39 thousand tons; on the other hand, the exports from Lower South America declined by 180 thousand tons. Although meat exports from West Africa declined, the exports from Sub-Saharan Africa increased to around 300 thousand tons in the mid-seventies, of which two-thirds were from Eastern and Southern Africa. Meat exports from Asia, which were largely from Mongolia, increased by 34 thousand tons, while those from North Africa/ Middle East declined slightly by 5 thousand tons over this period. (Table 3.1, Appendix A)

Between the early sixties and the mid-seventies, imports of meat rose in all the four regions, the largest increase being in North Africa/Middle East, where they were 2.8 times the 1961-65 level. In Asia the imports more than doubled to 422 thousand tons. These two regions shared four-fifths of the increase in meat imports into the Third World between the early sixties and the mid-seventies. Of the remaining increase, Latin America shared 93 thousand tons and Sub-Saharan Africa 13 thousand tons. The bulk of the meat imports into Asia were into South Asia in both the periods. In North Africa/Middle East, meat imports into Western Asia increased nearly 3.7 times over the same period. On the other hand, imports into West Africa declined slightly from 118 thousand to

30

112 thousand tons, although in Central Africa they increased one and a half times to 57 thousand tons. In Upper South America, the major meat importing sub-region in Latin America, meat imports rose from 91 thousand tons in 1961-65 to 163 thousand tons in 1973-77. This accounted for the bulk of the increase in meat imports into Latin America.

Of the projected supply-demand gap of 20.9 million tons in meat by 2000, North Africa/Middle East and Asia would together share nearly two-thirds. The projected deficit in Latin America would be relatively small, namely 2.7 million tons. Eighty-six percent of the shortfall of 4.7 million tons of meat in Sub-Saharan Africa would be from West Africa. East and Southeast Asia and Western Asia also would have large gaps of around 5 million tons each while the deficits in South Asia and Northern Africa would be of the order of 1.8 millions tons each. The gap in Upper South America would be 3.6 million tons. These are the areas where meat supplies need to be augmented.

Analysis of trade in meat classified by income levels of developing countries showed that the net trade surplus of high income countries declined from around 700 thousand tons in the early sixties to 125 thousand tons in the mid-seventies. The deficit of high income countries of Asia increased 131 thousand tons to 290 thousand tons over the same period. Countries of North Africa/Middle East in this group experienced a sudden jump in net imports from 21 thousand tons in 1961-65 to 272 thousand tons in 1973-77. The very low income countries of Sub-Saharan Africa were surplus in meat in both the periods, although the surplus declined slightly during the latter period. (Table 3.2, Appendix A)

Of the 20.9 million tons of supply-demand gap projected for 2000, the high income group of countries would account for less than 20 percent. The middle groups with income levels between $250 and $1,249 would share nearly two-thirds of the gap. The balance would be in very low-income countries.

Production and Consumption of Different Types of Meat

The composition of production and consumption of meat by type, namely, beef and buffalo meat, mutton and goat meat, pig meat and poultry meat during the early sixties and mid-seventies has been analyzed and the relevant data are presented in Tables 4.1 to 4.4, Appendix A. Whereas consumption has been projected by type of meat, output projections are in terms of total meat. Hence the estimates by meat type for 1990 and 2000 given in the tables relate to consumption only. The highlights are presented below.

Beef and Buffalo Meat

An estimated quantity of 11.6 million tons of beef and buffalo meat was produced in the 104 Third World countries in the mid-seventies. This represented 53 percent of the total meat, a decrease from 58 percent in the early sixties. This decline was reflected in consumption also from 55 to 50 percent. The fall in the relative importance of this type of meat was particularly

31

large in Latin America where its share in output decreased from 71 to 64 percent and in consumption from 68 to 61 percent. The growth in output of beef and buffalo meat at 2.2 percent per annum was only three-fourths of that in the total meat between the early sixties and the mid-seventies. By 2000, the share of this meat in the total meat consumption would go down further to 47 percent, though in Latin America it is projected to remain still high at nearly sixty percent. Of the 71.8 million tons of meat consumption projected for 2000, beef and buffalo meat would amount to 33.9 million tons.

Mutton and Goat Meat

Although the output of meat from small ruminants, namely sheep and goats, increased from 2.6 million tons in the early sixties to 3.0 million tons in the mid-seventies, its share of total meat declined from 16.4 percent to 13.6 percent. The decrease was particularly large in North Africa/Middle East from 48.9 to 41.8 percent over this period. In Latin America, the output of mutton and goat meat declined in absolute level also from 409 thousand tons to 388 thousand tons or at the rate of about half a percent a year. The fall in its relative share in total meat output is reflected also in the consumption share which declined from 16.9 percent to 14.2 percent. By 2000, the consumption of mutton and goat meat is projected to increase to 11.3 million tons, forming 16 percent of total meat consumption. In North Africa/Middle East, mutton and goat meat is the predominant type consumed; its share of 47 percent in the early sixties is projected to decrease to 37 percent by 2000. Even at this level the share of mutton and goat meat would be higher than that of beef and buffalo meat (32 percent) and poultry meat (31 percent) in this region.

Pig Meat

Pig meat is popular in Asia (33 percent) and Latin America (18 percent) and its consumption increased at 3.3 percent per annum which is marginally higher than the growth in the total meat consumption. The output of pig meat averaged 3.7 million tons in the mid-seventies as compared to 2.5 million tons in the early sixties. Pork consumption is projected to rise to 10.3 million tons by 2000 of which the shares of Latin America and Asia would be 4.9 million tons and 4.6 million tons respectively. The overall share of pork in 2000 may show a decline to 14 percent of the total meat consumption, compared to 17 percent in the early sixties.

Poultry Meat

Between the early sixties and the mid-seventies, the output of poultry meat expanded at the very rapid rate of 7.5 percent per annum, which is nearly two and a half times the growth of total meat production. The growth rate in Latin America was the fastest, followed by North Africa/Middle East. The share of poultry in the total meat output increased from 10 percent to 17 percent in the Third World, while in Latin America this share nearly doubled over this period. The growth in consumption was also rapid and was slightly higher than that of output, not only for the Third World countries as a whole but also in North Africa/Middle East and Asia. By 2000, the share of poultry in the total meat consumption is projected to reach 31 percent in North Africa/ Middle East and 23 percent for all the 104 study countries taken together.

Overall Position

The overall position of production and consumption of meat by type of meat is shown in Table 5.

Table 5. Distribution of Production and Consumption by Meat Type (percentage of total meat)

Meat type	1961-65	1973-77	1990	2000
Production				
Beef and Buffalo Meat	58	53	—	—
Mutton and Goat Meat	16	13	—	—
Pig meat	16	17	—	—
Poultry Meat	10	17	—	—
Total	100	100	—	—
Consumption				
Beef and Buffalo Meat	55	50	48	47
Mutton and Goat Meat	17	14	15	16
Pig meat	17	18	16	14
Poultry Meat	11	18	21	23
Total	100	100	100	100

Sources: Calculated from basic data in "Food and Agriculture Organization of the United Nations, "Global Agricultural Programming System Supply Utilization Accounts Tape," Rome, June 1980.

Food and Agriculture Organization of the United Nations: FAO Production Yearbook Tapes 1975 and 1979, Rome 1976 and 1980 and IFPRI Projections.

Trends and Projections of Feed Use of Cereals[7]

The domestic utilization of cereals for animal feed in the Third World expanded faster than its output of meat. Between 1961-65 and 1973-77, cereal feed use increased from around 35 million tons to 63 million tons or at an annual rate of around 5.1 percent. This compares with the growth of 2.9 percent in meat production over the same period. The growth in feed use was also faster than that in direct food use of these grains which rose at 2.8 percent. Expansion in animal feed has been rapid particularly in regions and sub-regions in which the output of poultry meat also increased fast, especially in Latin America. At the sub-regional level, cereal feed use more than doubled in East and Southeast Asia and in Northern Africa and trebled in Central America and the Caribbean. Half of the domestic utilization of cereals for feed in the Third World was in Latin America. (Table 5.1, Appendix A)

[7]Excludes rice bran.

Projections of Feed Use to 1990 and 2000

Tentative estimates of cereal feed utilization in 1990 and 2000 worked out as part of IFPRI's global food gap analysis indicate that the total requirements of these grains for feed might increase to 137 million tons in 1990 and to nearly 230 million tons by the end of the century.[8] (Table 5.2, Appendix A) These estimates employed the income elasticity of demand for meat as a proxy for that of animal feed and hence are essentially based on the projected growth in meat consumption. Thirty-five percent of the increase between 1977 and 2000 would be in Latin America and another 35 percent in North Africa/Middle East. The share of Latin America in the Third World feed use would decline from around 50 percent in 1977 to 40 percent by 2000 while that of North Africa/Middle East would go up to one-third from about a quarter over the same period. IFPRI is re-examining these feed estimates and the indications are that they may need an upward revision.

Even at this level, feed use of cereals in the Third World countries would account for a quarter of their total domestic utilization in 2000. This share would vary from one-eighth in Asia to nearly half of total domestic utilization in Latin America.

Past Trends and Projections
of Supply-Demand Balances of Cereals

Trends in Production

Production of cereal grains in the Third World countries (excluding China) increased from 258 million tons in the early sixties to 361 million tons in the mid-seventies or at an annual rate of 2.8 percent per annum. (Table 6.1, Appendix A) This rate is higher than the population growth of 2.6 percent. Fifty-six percent of the increase in production was from Asia, although the most rapid growth was in Latin America. The increase in Sub-Saharan Africa was the smallest in both relative and absolute terms; it was also lower than the population growth and thus per capita output of cereals declined by more than one percent per annum in this region.

Projections for 1990 and 2000

If past trends continue, the output of cereals in the Third World is projected to increase to 576 million tons in 1990 and further to 803 million tons by 2000. The average growth rate is projected at 3.3 percent - an acceleration of 0.5 percent over the 1961-77 growth. Regional differences would persist, ranging from 2.7 percent in Sub-Saharan Africa to 3.8 percent in Latin America. Asia would still account for a little over half of the Third World's cereal output in 2000 while Latin America's share would be a little less than a quarter. At the

---

[8]IFPRI is currently undertaking a detailed analysis of livestock feed. For a fuller discussion of the past trends and projections to the future in feed use of major food crops see Leonardo A. Paulino, "Food in the Third World: Past Trends and Projections to 2000," International Food Policy Research Institute, Washington, D.C., 1984. (Mimeographed.)

sub-regional level, the rate of growth in Upper South America (4.3 percent) would be more than three times that in West Africa (1.4 percent). (Table 6.2, Appendix A)

Domestic Utilization and Trade

Between the early sixties and the mid-seventies, domestic utilization of cereals for direct food use increased at about the same rate as that of their output in the Third World as a whole. The growth in consumption in North Africa/Middle East was 50 percent higher than that in output, thus necessitating large imports into the region. In Sub-Saharan Africa, poor production performance led to a difference between the two growth rates of almost 1.0 percent per annum. Hence the net imports of cereals into both the regions trebled over the period. The overall net imports of cereals into the Third World countries increased threefold from 9.8 million tons in the early sixties to 28.5 million tons in the mid-seventies. Half of the latter-period imports were into Asia. The surplus in Latin America in the mid-seventies was nearly one-tenth of that in the early sixties. (Tables 6.3 and 6.6, Appendix A)

Projected Demand for Cereals: 1990 and 2000

The demand for cereals was projected separately for direct human consumption, livestock feed, requirements for seed and allowances for wastage and other uses. If past trends in income of developing countries continue into the future, the domestic demand for cereals for direct consumption is expected to rise to 422 million tons by 1990 and 528 million tons by 2000 at the average rates of 2.8 and 2.6 percent per annum, respectively. (Table 6.4, Appendix A). Since the population growth is projected to slow down to 2.4 percent over the 1977-2000 period, the per capita consumption of cereals as direct food would show only a small increase by the end of the century. On the other hand, the derived demand for cereals as livestock feed would go up faster as was seen in Section 7. Total demand for cereals would go up to 639 million tons in 1990 and 869 million tons in 2000, including the requirements for seed and allowance for wastage. The distribution of this demand by sub-regions and regions is presented in Table 6.5, Appendix A.

Projected Cereal Gaps

Thus if past trends of production and per capita incomes in developing countries continue into the future, food projections point to an overall net deficit of about 66 million tons of cereals in the Third World by the end of the century. This gap represents around 8 percent of the projected demand for cereals. However, it is nearly 2.2 times the net trade deficit in the mid-seventies. Both Asia and Latin America would be surplus in cereals, but North Africa/Middle East and Sub-Saharan Africa will have net deficits amounting to 66 and 28 million tons respectively. The deficits will be particularly large in West Africa, Upper South America and both the sub-regions of North Africa/ Middle East. (Table 6.7, Appendix A)

Policy Implications of the
Meat and Cereal Supply-Utilization Trends

The analysis of supply-demand balances for meat and cereals in the preceding sections shows that the Third World countries would be facing large deficits in both meat and cereals if past trends in production and per capita income continue into 2000. The feed component of the demand for cereals would increase faster than the direct demand for human consumption, largely because of the relatively rapid growth in the demand for livestock products as incomes rise and, partly, of the expected changes in the structure of livestock production. The rapidly increasing derived demand for feed could increasingly draw resources away from the production of the basic staples that are used directly for food, to the disadvantage of the poor. The analysis also indicates that the areas that are likely to face large deficits in meat would be located in Western Asia, East and Southeast Asia, West Africa and Upper South America.

If the projected demand were to be met entirely from domestic production, the projected growth rate of meat of 3.4 percent per annum over the period 1977-2000 would have to be raised by 50 percent. In North Africa/Middle East, a near doubling of the growth rate would be necessary to achieve comparable self-sufficiency.

Under the alternative assumption of slower income growth, that is 25 percent lower than the 1966-77 trend, the gap between projected production and consumption would still be large, at 10 million tons by 2000, representing 16 percent of the estimated consumption. To meet this deficit from domestic supplies, production growth has to be accelerated by 25 percent over the projected 1977-2000 growth rate. Further even under this alternative, except Latin America, the three other regions would be deficit in meat.

Some of these trends are already evident from the available data on trade in livestock and poultry products for more recent years. Developing countries (excluding China) which were net exporters of meat until the mid-seventies became net importers in the latter half of the decade and by 1981, their net imports rose to 1.16 million tons of all types of meat inclusive of offals, and this trend is likely to continue.

As transfers of livestock products on a large scale from developed to developing countries in the form of either trade or aid are less likely, considerable efforts are needed to accelerate their production with a view to reduce the large gaps between the projected production and demand. Otherwise relative prices would rise or there would be shifts in consumption which in some cases would cause considerable hardships to the poor consumers in some of the Third World countries. The demand growth of livestock products is more rapid in urban areas.

It may be pointed out that whereas the projections of demand for meat are made by type of meat, those of supply are based on the past trends in the total meat output. The scope for increasing the production of individual types of meat depends not only on climatic and environmental factors, type of farming and the physical resources in each area, but also on the availability of inputs and infrastructure facilities. Where the major input of livestock production, namely feed, is inadequate to support the output expansion required to meet the anticipated demand, consideration would have to be given to the scope for increasing the domestic output of feed; and where this is not possible, recourse could be had to imports if foreign exchange is not a constraint.

Acceleration of the development of poultry and pigmeat production would lead to greater derived demand for coarse grains as livestock feed. In several developing countries in all the regions maize and other coarse grains are directly consumed as food and their increased feed use would imply a diversion from their food use. Competition between the two uses may result in higher prices which would cause hardship to the vulnerable sections of the population. A food-feed competition regarding the use of grains may develop not only between the rich and the poor people within the country but also between the richer and poorer countries. Action would have to be taken in several directions to meet the situation. The productivity of cereal feeds needs to be raised wherever potential exists. Greater emphasis should be laid on research and technology to improve yields of feedgrains and the feeding efficiency.

In some areas where the food problem is more severe, and where scope exists for development of small ruminant production based on the utilization of grasses and farm by-products, steps would have to be taken to encourage this. There is no doubt, however—and the demand and supply projections for the livestock and poultry products indicate this—that over large areas demand for feeds and fodder would increase very rapidly in the coming decades, resulting in increased competition for land. More intensive use of land would have to be resorted to and more attention than at present would have to be given to research and development of feed resources within the country, especially in countries experiencing foreign exchange difficulties. Greater efforts are required for development of new sources of feed, greater use of by-products and agricultural wastes and compound feeds.

The potentials offered by isolated soy protein in helping ease the problems likely to be posed by the projected demand-supply imbalances of meat in the Third World need to be examined. As a food ingredient that can augment the supply of meat products in developing countries where the growth of meat demand would greatly outpace increases in meat output and, consequently, the production of grains for livestock feed may draw resources away from the production of grains for direct human consumption, isolated soy protein can help relieve the resulting pressure on the prices of basic food staples. Studies need to be made, however, on the overall costs and benefits of utilizing the commodity in these countries.

Appendix A

Supplementary Tables

1. Production
   1.1 Production and growth rate of total meat, 1961/65 and 1973/77 averages with projections to years 1990 and 2000, grouped by geographical region and sub-region.

   1.2 Production and growth rate of total meat, 1961/65 and 1973/77 averages with projections to years 1990 and 2000, grouped by 1961-77 trend value of real GNP per capita.

2. Consumption
   2.1 Consumption and growth rate of total meat, 1961/65 and 1973/77 averages with projections to years 1990 and 2000, grouped by geographical region and sub-region.

   2.2 Consumption and growth rate of total meat, 1961/65 and 1973/77 averages with projections to years 1990 and 2000, grouped by 1961-77 trend value of real GNP per capita.

   2.3 Projections of meat consumption and surpluses/deficits in 1990 and 2000 under low income growth assumption.

3. Trade, Surpluses and Deficits
   3.1 Imports, exports and net trade in total meat, 1961/65 and 1973-77 averages with the surplus and deficit projections to years 1990 and 2000, grouped by geographical region and sub-region.

   3.2 Imports, exports and net trade in total meat, 1961/65 and 1973-77 averages with the surplus and deficit projections to years 1990 and 2000, grouped by 1961-77 trend value of real GNP per capita.

4. Production and Consumption by Type of Meat
   4.1 Production, growth rate and percentage share in total meat of beef and veal, 1961/65 and 1973/77 averages.

   4.2 Consumption, growth rate and percentage share in total meat of beef and veal, 1961/65 and 1973/77 averages with projections to years 1990 and 2000.

   4.3 Production, growth rate and percentage share in total meat of mutton and lamb, 1961/65 and 1973/77 averages.

   4.4 Consumption, growth rate and percentage share in total meat of mutton and lamb, 1961/65 and 1973/77 averages with projections to years 1990 and 2000.

4.5 Production, growth rate and percentage share in total meat of pig meat, 1961/65 and 1973/77 averages.

4.6 Consumption, growth rate and percentage share in total meat of pig meat, 1961/65 and 1973/77 averages with projections to years 1990 and 2000.

4.7 Production, growth rate and percentage share in total meat of poultry meat, 1961/65 and 1973/77 averages.

4.8 Consumption, growth rate and percentage share in total meat of poultry meat, 1961/65 and 1973/77 averages with projections to years 1990 and 2000.

5. Utilization of Grains for Feed

5.1 Trends in domestic utilization of grains for feed in developing countries, 1961/65 and 1973/77 averages, grouped by geographical region and sub-region.

5.2 Domestic utilization of grains for feed in developing countries, trend value 1977 and projections to years 1990 and 2000, grouped by geographical region and sub-region

6. Production of Grains

6.1 Trends in production of grains in developing countries, 1961/65 and 1973/77 averages

6.2 Production of grains in developing countries, trend value 1977 and projections to years 1990 and 2000, grouped by geographical region and sub-region

6.3 Trends in domestic utilization of grains for food in developing countries, 1961/65 and 1973/77 averages

6.4 Domestic utilization of grains for food in developing countries, trend value 1977 and projections to years 1990 and 2000, grouped by geographical region and sub-region

6.5 Total domestic utilization of grains in developing countries, trend value 1977 and projections to 1990 and 2000

6.6 Trends in exports, imports and net trade of grains in developing countries, 1961/65 and 1973/77 averages

6.7 Net surplus deficit of grains, in developing countries trend value 1977 and projections to 1990 and 2000

Table 1.1 Production and growth rate of total meat, 1961/65 and 1973/77 averages with projections to years 1990 and 2000, grouped by geographical region and sub-region

('000 mt)

Region/Sub-Region	Average Production			Production Projections				
				Trend Value			Growth Rate	
	1961/65	1973/77	Growth Rate 1961/65-1973/77 (percent)	1977	1990	2000	1977-1990 (percent)	1977-2000 (percent)
Total 104 countries	15,578.6	22,019.0	2.93	23,484.7	35,782.3	50,924.6	3.29	3.42
Asia	3,422.1	4,758.0	2.78	5,062.9	7,518.8	10,532.9	3.09	3.24
South Asia	1,279.8	1,593.6	1.84	1,655.7	2,142.1	2,642.5	2.00	2.05
East and Southeast Asia	2,142.3	3,164.4	3.30	3,407.2	5,376.7	7,890.4	3.57	3.72
North Africa/Middle East	1,911.1	2,875.9	3.46	3,069.4	5,010.0	7,501.6	3.84	3.96
Northern Africa	695.6	1,060.0	3.57	1,150.6	1,879.3	2,843.5	3.85	4.01
Western Asia	1,215.5	1,815.9	3.40	1,918.8	3,130.7	4,658.1	3.84	3.93
Sub-Saharan Africa	2,181.1	2,874.9	2.33	3,090.3	4,505.1	6,257.1	2.94	3.11
West Africa	714.9	861.0	1.56	923.5	1,219.7	1,569.4	2.16	2.33
Central Africa	154.3	269.8	4.77	297.6	549.6	887.2	4.83	4.86
Eastern and Southern Africa	1,311.9	1,744.1	2.40	1,869.2	2,735.8	3,800.5	2.97	3.13
Latin America	8,064.3	11,510.2	3.01	12,262.1	18,748.4	26,633.0	3.32	3.43
Central America and Caribbean	1,374.9	2,153.4	3.81	2,328.7	3,858.3	5,745.0	3.96	4.00
Upper South America	3,403.2	5,472.4	4.04	5,934.1	10,124.6	15,397.7	4.20	4.23
Lower South America	3,286.2	3,884.4	1.40	3,999.3	4,765.5	5,490.3	1.36	1.39

Source: Calculated from basic data in Food and Agriculture Organization of the United Nations, FAO Production Yearbook Tapes 1975 and 1979, Rome 1976 and 1980, and IFPRI Projections.

Table 1.2 Production and growth rate of total meat, 1961/65 and 1973/77 averages with projections to years 1990 and 2000, grouped by 1961-77 trend value of real GNP per capita

('000 mt)

Region/1977 GNP Per Capita Levels	Average Production			Production Projections				
				Trend Value			Growth Rate	
	1961/65	1973/77	Growth Rate 1961/65-1973/77 (percent)	1977	1990	2000	1977-1990 (percent)	1977-2000 (percent)
Total 104 Countries	15,578.6	22,019.0	2.93	23,484.7	35,782.3	50,924.6	3.29	3.42
Less than $250	3,257.7	4,084.8	1.90	4,335.8	5,900.0	7,664.3	2.40	2.51
Asia	1,858.5	2,459.3	2.36	2,601.5	3,622.5	4,740.6	2.58	2.64
North Africa/ Middle East	170.5	170.9	0.02	178.9	237.3	295.0	2.20	2.20
Sub-Saharan Africa	1,192.8	1,405.5	1.38	1,502.9	1,966.4	2,532.7	2.09	2.30
Latin America	35.9	49.1	2.64	52.5	73.8	96.0	2.65	2.66
$250-$499	2,362.6	3,402.7	3.09	3,652.4	5,573.0	7,839.0	3.30	3.38
Asia	1,044.2	1,474.1	2.91	1,570.3	2,295.6	3,106.1	2.96	3.01
North Africa/ Middle East	487.8	704.0	3.10	764.4	1,194.2	1,699.7	3.49	3.54
Sub-Saharan Africa	721.6	1,041.8	3.11	1,121.6	1,738.7	2,500.3	3.43	3.55
Latin America	109.0	182.8	4.40	196.1	344.5	532.9	4.43	4.44
$500-$1,249	2,908.3	4,256.3	3.22	4,534.7	7,135.3	10,431.8	3.55	3.69
Asia	438.8	697.6	3.94	749.3	1,350.4	2,297.3	4.64	4.99
North Africa/ Middle East	888.9	1,296.4	3.19	1,368.1	2,077.0	2,895.4	3.26	3.31
Sub-Saharan Africa	260.0	419.1	4.06	456.6	787.3	1,206.9	4.28	4.32
Latin America	1,320.6	1,843.2	2.82	1,960.7	2,920.6	4,032.2	3.11	3.18
$1,250 and over	7,050.0	10,275.2	3.19	10,961.8	17,174.0	24,989.5	3.51	3.65
Asia	80.6	127.0	3.86	141.8	250.3	388.9	4.47	4.48
North Africa/ Middle East	363.9	704.6	5.66	758.0	1,501.5	2,611.5	5.40	5.53
Sub-Saharan Africa	6.7	8.5	2.00	9.2	12.7	17.2	2.51	2.76
Latin America	6,598.8	9,435.1	3.02	10,052.8	15,409.5	21,971.9	3.34	3.46

Source: Calculated from basic data in Food and Agriculture Organization of the United Nations, FAO Production Yearbook Tapes 1975 and 1979, Rome 1976 and 1980, and IFPRI Projections.

Table 2.1 Consumption and growth rate of total meat, 1961/65 and 1973/77 averages with projections to years 1990 and 2000, grouped by geographical region and sub-region
('000 mt)

Region/Sub-Region	Average Consumption			Consumption Projections				
	1961/65	1973/77	Growth Rate 1961/65-1973/77 (percent)	1977°	1990	2000	Growth Rate 1977-1990 (percent)	1977-2000 (percent)
Total '04 Countries	14,934.6	21,689.7	3.16	23,208.9	43,818.1	71,788.2	5.01	5.03
Asia	3,518.8	4,986.1	2.95	5,247.5	10,254.3	17,233.1	5.29	5.31
South Asia	1,290.3	1,590.6	1.76	1,654.7	2,934.4	4,412.2	4.51	4.36
East and Southeast Asia	2,228.5	3,395.5	3.57	3,592.6	7,319.9	12,820.9	5.63	5.69
North Africa/Middle East	2,034.8	3,285.3	4.07	3,625.9	8,045.9	14,307.5	6.32	6.15
Northern Africa	761.6	1,131.0	3.35	1,209.2	2,598.3	4,597.1	6.06	5.98
Western Asia	1,273.2	2,154.3	4.48	2,416.7	5,447.6	9,710.4	6.45	6.23
Sub-Saharan Africa	2,143.6	2,734.0	2.05	2,871.9	5,888.7	10,987.7	5.68	6.01
West Africa	734.2	896.7	1.68	926.6	2,443.9	5,621.7	8.42	8.15
Central Africa	189.8	316.2	4.35	349.0	596.3	893.2	4.21	4.17
Eastern and Southern Africa	1,219.6	1,521.1	1.86	1,596.3	2,848.7	4,472.8	4.57	4.58
Latin America	7,237.4	10,684.3	3.30	11,463.6	19,629.2	29,259.9	4.22	4.16
Central America and Caribbean	1,274.9	2,046.4	4.02	2,296.4	3,835.9	5,753.6	4.03	4.07
Upper South America	3,342.1	5,354.7	4.01	5,849.5	11,767.8	19,007.1	5.52	5.26
Lower South America	2,620.4	3,283.2	1.90	3,317.7	4,025.5	4,499.2	1.50	1.33

° Consumption data for 1977 refer to trend values.

Sources: Calculated from basic data in Food and Agriculture Organization of the United Nations "Global Agricultural Programming System Supply Utilization Accounts Tape." Rome, June 1980 and IFPRI Projections.

Table 2.2 Consumption and growth rate of total meat, 1961/65 and 1973/77 averages with projections to years 1990 and 2000, grouped by 1961–77 trend value of real GNP per capita

('000 mt)

Region/1977 GNP Per Capita Levels	Average Consumption			Consumption Projections				
	1961/65	1973/77	Growth Rate 1961/65-1973/77 (percent)	1977°	1990	2000	Growth Rate 1977-1990 (percent)	1977-2000 (percent)
Total 104 Countries	14,934.6	21,689.7	3.16	23,208.9	43,818.1	71,788.2	5.01	5.03
Less than $250	3,204.9	4,015.7	1.90	4,169.8	6,975.2	10,195.1	4.04	3.96
Asia	1,861.3	2,458.8	2.35	2,575.0	4,321.7	6,258.6	4.06	3.94
North Africa/Middle East	169.5	171.7	0.11	166.7	257.4	351.4	3.40	3.30
Sub-Saharan Africa	1,139.1	1,336.9	1.34	1,377.8	2,312.2	3,458.4	4.06	4.08
Latin America	35.0	48.3	2.72	50.3	83.9	126.7	4.01	4.10
$250-499	2,436.4	3,343.2	2.67	3,494.0	8,131.1	16,251.7	6.71	6.91
Asia	1,038.1	1,442.5	2.78	1,489.4	3,643.0	7,327.9	7.12	7.17
North Africa/Middle East	527.3	744.4	2.92	784.7	1,448.5	2,284.9	4.83	4.76
Sub-Saharan Africa	773.9	1,002.3	2.18	1,052.2	2,741.8	6,178.3	7.65	8.00
Latin America	97.1	154.0	3.92	167.7	297.8	460.6	4.51	4.49
$500-$1,249	2,836.0	4,130.3	3.18	4,453.8	9,331.9	16,278.9	5.85	5.80
Asia	411.9	679.2	4.26	744.2	1,590.1	2,746.3	6.01	5.84
North Africa/Middle East	942.9	1,345.7	3.01	1,426.7	3,579.8	7,007.6	7.33	7.17
Sub-Saharan Africa	218.0	368.5	4.47	412.5	783.4	1,266.5	5.06	5.00
Latin America	1,263.2	1,736.9	2.69	1,870.4	3,378.6	5,258.5	4.65	4.60
$1,250 and Over	6,457.3	10,200.5	3.88	11,091.3	19,379.9	29,062.5	4.39	4.28
Asia	207.5	405.6	5.74	438.9	699.5	900.3	3.65	3.17
North Africa/Middle East	395.1	1,023.5	8.26	1,247.8	2,760.2	4,663.6	6.30	5.90
Sub-Saharan Africa	12.6	26.3	6.32	29.4	51.3	84.5	4.38	4.70
Latin America	5,842.1	8,745.1	3.42	9,375.2	15,868.9	23,414.1	4.13	4.06

° Consumption data for 1977 refer to trend values.

Sources: Calculated from basic data in Food and Agriculture Organization of the United Nations "Global Agricultural Programming System Supply Utilization Accounts Tape." Rome, June 1980 and IFPRI Projections.

Table 2.3 Projections of meat consumption and surpluses/deficits in 1990 and 2000 under low income growth assumption.

('000 mt)

Region /Sub-Region	Consumption under low income growth assumption		Surplus/Deficit	
	1990	2000	1990	2000
Total 104 Study Countries	39,653.7	61,168.0	(3,871.4)	(10,223.0)
Asia				
South Asia	9,347.4	14,208.6	(1,828.6)	(3,675.7)
East and Southeast Africa	2,774.1	3,983.0	(632.0)	(1,340.5)
	6,573.3	10,225.6	(1,196.6)	(2,335.2)
North Africa/Middle East	7,216.9	11,720.3	(2,206.9)	(4,218.7)
Northern Africa	2,349.8	3,805.6	(470.5)	(962.1)
Western Asia	4,867.1	7,914.7	(1,736.4)	(3,256.6)
Sub-Saharan Africa	4,633.7	9,071.8	(128.6)	(2,814.3)
West Africa	2,109.1	4,156.7	(889.4)	(2,586.9)
Central Africa	570.5	817.0	(20.9)	70.2
Eastern and Southern Africa	1,954.1	4,098.1	781.7	(297.6)
Latin America	18,455.7	26,167.3	292.7	465.7
Central America and Caribbean	3,669.1	5,306.0	189.2	459.0
Upper South America	10,800.4	16,425.9	(675.8)	(1,028.2)
Lower South America	3,986.2	4,435.4	779.3	1,054.9

Sources: Calculated from basic data in Food and Agriculture Organization of the United Nations, "Global Agricultural Programming System Supply Utilization Accounts Tape." Rome, June 1980 and IFPRI Projections.

Note: Figures in brackets refer to net deficits.

Table 3.1 Imports, exports and net trade in total meat, 1961/65 and 1973/77 averages with surplus and deficit projections to years 1990 and 2000, grouped by geographical region and sub-region

('000 mt)

Region/Sub-Region	1961/65			1973/77			1977[a]			Projections					
										1990			2000		
	Imports	Exports	Net	Imports	Exports	Net	Surplus	Deficit	Net	Surplus	Deficit	Net	Surplus	Deficit	Net
Total 104 Countries	735.6	1,495.4	759.8	1,314.6	1,618.9	304.3	1,662.9	1,387.1	275.8	1,759.5	9,796.0	(8,036.5)	3,715.3	24,579.9	(20,864.6)
Asia	204.7	101.0	(103.7)	421.8	135.4	286.4	203.7	388.3	(184.6)	196.6	2,932.2	(2,735.6)	642.1	7,342.4	(6,700.3)
South Asia	8.5	1.9	(6.6)	8.8	3.3	(5.5)	19.1	18.1	1.0	0	792.3	(792.3)	0	1,769.7	(1,769.7)
East and Southeast Asia	196.2	99.1	(97.1)	413.0	132.1	(280.9)	184.6	370.2	(185.6)	196.6	2,139.9	(1,943.3)	642.1	5,572.7	(4,930.6)
North Africa/ Middle East	142.2	42.6	(99.6)	398.0	37.7	(360.3)	39.5	596.0	(556.5)	60.8	3,096.7	(3,035.9)	276.8	7,082.8	(6,306.0)
Northern Africa	57.7	6.1	(51.6)	86.1	11.1	(75.0)	27.3	85.9	(58.6)	60.8	779.8	(719.0)	276.8	2,030.5	(1,753.7)
Western Asia	84.5	36.5	(48.0)	311.9	26.6	(285.3)	12.2	510.1	(497.9)	0	2,316.9	(2,316.0)	0	5,052.3	(5,052.3)
Sub-Saharan Africa	195.3	263.4	68.1	208.7	301.2	92.5	403.1	184.7	218.4	511.5	1,895.3	(1,383.8)	903.1	5,633.9	(4,730.8)
West Africa	117.8	112.0	(5.8)	112.4	88.9	(23.5)	95.5	98.6	(3.1)	58.8	1,283.0	(1,224.2)	120.2	4,172.7	(4,052.5)
Central Africa	39.4	5.3	(34.1)	56.8	10.1	(46.7)	6.5	57.9	(51.4)	41.9	88.6	(46.7)	109.0	115.0	(6.0)
Eastern and Southern Africa	38.1	146.1	108.0	39.5	202.2	162.7	301.1	28.2	272.9	410.8	523.7	(112.9)	673.9	1,346.2	(672.3)
Latin America	193.4	1,088.4	895.0	286.1	1,144.6	858.5	1,018.6	255.1	798.5	990.6	1,871.8	(881.2)	1,893.3	4,520.8	(2,627.5)
Central America and Caribbean	57.4	170.6	113.2	98.9	209.0	110.1	133.3	101.0	32.3	237.3	215.1	22.2	510.6	519.5	(8.9)
Upper South America	90.9	143.9	53.0	163.3	342.1	178.8	188.8	104.2	84.6	13.3	1,656.6	(1,643.3)	391.6	4,001.2	(3,609.6)
Lower South America	45.1	773.9	728.8	23.9	593.5	569.6	694.5	12.9	681.6	740.0	0.1	739.9	991.1	0.1	991.0

Note: Figures in brackets are net imports or net deficits.

[a] Data for 1977 refer to trend values.

Sources: Calculated from basic data in Food and Agriculture Organization of the United Nations "Global Agricultural Programming System Supply Utilization Accounts Tape." Rome, June 1980, and IFPRI Projections.

Table 3.2 Imports, exports and net trade in total meat, 1961/65 and 1973/77 averages with surplus and deficit projections to years 1990 and 2000, grouped by 1961–77 trend value of real GNP per capita

('000 mt)

Region/1977 GNP Per Capita Levels	1961/65			1973/77			1977a			Projections 1990			2000		
	Imports	Exports	Net	Imports	Exports	Net	Surplus	Deficit	Net	Surplus	Deficit	Net	Surplus	Deficit	Net
Total 104 Countries	735.6	1,495.4	759.8	1,314.6	1,618.9	304.3	1,662.9	1,387.1	275.8	1,759.5	9,796.0	(8,035.8)	3,715.3	24,579.9	(20,863.6)
Less than $250	60.2	147.6	87.4	69.4	143.3	73.9	233.1	67.1	166.0	246.7	1,321.9	(1,075.2)	532.5	3,063.3	(2,530.8)
Asia	9.5	10.6	1.1	12.6	5.7	(6.9)	44.6	18.1	26.5	93.1	792.3	(699.2)	251.7	1,769.7	(1,518.0)
North Africa and Middle East	0	0	0	0	0	0	0	0	12.2	0	20.1	(20.1)	0	56.4	(56.4)
Sub-Saharan Africa	50.5	135.8	85.3	56.4	136.5	80.1	174.1	49.0	125.1	153.6	499.4	(345.8)	280.8	1,260.5	(925.7)
Latin America	0.2	1.2	1.0	0.4	1.1	0.7	2.2	0	2.2	0	10.1	(10.1)	0	30.7	(30.7)
$250–$499	150.2	96.9	(53.3)	142.8	137.7	(5.1)	265.9	107.5	158.4	302.5	2,860.8	(2,558.1)	603.0	9,015.9	(8,412.7)
Asia	17.7	22.0	4.3	11.1	17.7	6.6	95.7	14.8	80.9	0	1,347.4	(1,347.4)	0	4,221.8	(4,221.8)
North Africa and Middle East	27.5	4.8	(22.7)	46.3	11.0	(35.3)	15.7	36.0	(20.3)	60.8	315.1	(254.3)	151.2	736.4	(585.2)
Sub-Saharan Africa	101.1	57.8	(43.3)	78.9	75.0	(3.9)	126.1	56.7	69.4	195.0	1,198.3	(1,003.1)	379.5	4,057.7	(3,678.0)
Latin America	3.9	12.3	8.4	6.5	34.0	27.5	28.4	0	28.4	46.7	0	46.7	72.3	0	72.3
$500–$1,249	222.7	252.1	29.4	281.4	392.1	110.7	274.2	193.3	80.9	401.5	2,598.3	(2,196.6)	901.9	6,749.2	(5,847.1)
Asia	32.1	53.8	21.7	81.2	87.4	6.2	63.4	58.3	5.1	103.5	343.3	(239.7)	390.4	839.5	(449.0)
North Africa and Middle East	86.9	31.1	(55.8)	78.2	24.8	(53.4)	11.6	70.2	(58.6)	0	1,502.8	(1,502.8)	0	4,112.2	(4,112.2)
Sub-Saharan Africa	37.7	69.8	32.1	55.4	89.7	34.3	102.9	58.8	44.1	162.9	159.0	3.9	242.8	302.4	(59.6)
Latin America	66.0	97.4	31.4	66.6	190.2	123.6	96.3	6.0	90.3	135.1	593.2	(458.0)	268.7	1,495.1	(1,226.3)
$1,250 and over	302.5	998.8	696.3	821.0	945.8	124.8	889.7	1,019.2	(129.5)	808.8	3,015.0	(2,205.9)	1,677.9	5,751.5	(4,073.0)
Asia	145.4	14.6	(130.8)	316.9	24.6	(292.3)	0	297.1	(297.1)	0	449.2	(449.2)	0	511.4	(511.4)
North Africa and Middle East	27.8	6.7	(21.1)	273.5	1.9	(271.6)	0	489.8	(489.8)	0	1,258.7	(1,258.7)	125.6	2,177.8	(2,052.1)
Sub-Saharan Africa	6.0	0	(6.0)	18.0	0	(18.0)	0	20.2	(20.2)	0	38.6	(38.6)	0	67.3	(67.3)
Latin America	123.3	977.5	854.2	212.6	919.3	706.7	889.7	212.1	277.6	808.8	1,268.5	(459.4)	1,552.3	2,995.0	(1,442.2)

Note: Figures in brackets are net imports or net deficits.

a Data for 1977 refer to trend values.

Sources: Calculated from basic data in Food and Agriculture Organization of the United Nations "Global Agricultural Programming System Supply Utilization Accounts Tape." Rome, June 1980 and IFPRI Projections.

Table 4.1 Production, growth rate and percentage share in total meat of beef and veal[a], 1961/65 and 1973/77 averages.
('000 mt)

Region	Average Production			% of Beef & Veal in Total Meat	
	1961/65	1973/77	Growth Rate 1961/65-1973/77	1961/65	1973/77
			(percent)		
Total 104 Countries	8,982.7	11,635.0	2.18	57.51	52.84
Asia	1,217.6	1,591.2	2.26	35.58	33.44
North Africa and Middle East	731.8	1,030.5	2.89	38.29	35.83
Sub-Saharan Africa	1,327.0	1,664.2	1.91	59.75	57.88
Latin America	5,706.3	7,349.1	2.13	70.76	63.85

[a] Includes buffalo meat.

Source: Food and Agriculture Organization of the United Nations: FAO Production Yearbook tapes 1975 and 1979, Rome, 1976 and 1980.

Table 4.2 Consumption, growth rate and percentage share in total meat of beef and veal[a], 1961/65 and 1973/77 averages with projections to years 1990 and 2000.
('000 mt)

Region	Average Consumption					Consumption Projections							
	1961/65	% of Total Meat	1973/77	% of Total Meat	Growth Rate 1961/65-1973/77	1977[b]	% of Total Meat	1990	% of Total Meat	2000	% of Total Meat	Growth Rate 77-90	77-2000
					(percent)							(percent)	(percent)
Total 104 Countries	8,240.6	55.18	10,903.2	50.27	2.36	11,345.5	49	20,990.8	48	33,895.5	47	4.85	4.87
Asia	1,254.8	35.66	1,640.4	32.90	2.26	1,701.7	32	3,418.5	33	5,977.1	35	5.51	5.61
North Africa/Middle East	807.1	39.66	1,170.8	35.64	3.15	1,250.0	35	2,642.2	33	4,541.3	32	5.93	5.77
Sub-Saharan Africa	1,252.5	58.43	1,531.9	56.04	1.69	1,590.3	55	3,269.0	56	6,070.4	55	5.70	6.00
Latin America	4,926.2	68.06	6,560.0	61.40	2.42	6,803.5	60	11,661.1	60	17,306.7	59	4.23	4.14

[a] Includes buffalo meat.

[b] Data for 1977 refer to trend values.

Sources: Calculated from basic data in Food and Agriculture Organization of the United Nations, "Global Agricultural Programming System Supply Utilization Accounts Tape," Rome, June, 1980, and IFPRI Projections.

Table 4.3 Production, growth rate and percentage share in total meat of mutton and lamb°, 1961/65 and 1973/77 averages.

('000 mt)

Region	Production			% of Mutton & Lamb in Total Meat	
	1961/65	1973/77	Growth Rate 1961/65-1973/77 (percent)	1961/65	1973/77
Total 104 Countries	2,563.2	2,997.0	1.31	16.41	13.61
Asia	678.8	815.9	1.54	19.84	17.15
North Africa and Middle East	933.8	1,202.3	2.13	48.86	41.80
Sub-Saharan Africa	541.2	590.9	0.73	24.37	20.55
Latin America	409.3	387.9	-0.45	5.08	3.37

° Includes goat meat.

Source: Food and Agriculture Organization of the United Nations: FAO Production Yearbook tapes 1975 and 1979. Rome, 1976 and 1980.

Table 4.4 Consumption, growth rate and percentage share in total meat of mutton and lamb °, 1961/65 and 1973/77 averages with projection to years 1990 and 2000.

('000 mt)

Region	Average Consumption					Consumption Projections							
	1961/65	% of Total Meat	1973/77	% of Total Meat	Growth Rate 1961/65-1973/77 (percent)	1977ᵇ	% of Total Meat	1990	% of Total Meat	2000	% of Total Meat	Growth Rate 77-90 (percent)	77-2000 (percent)
Total 104 Countries	2,525.6	16.91	3,078.8	14.20	1.66	3,196.6	14	6,459.6	15	11,304.2	16	5.56	5.65
Asia	656.7	18.66	796.8	15.98	1.62	832.1	16	1,640.8	16	2,819.5	16	5.36	5.45
North Africa/ Middle East	965.9	47.47	1,337.1	40.70	2.75	1,418.5	38	3,044.5	38	5,333.9	37	6.05	5.93
Sub-Saharan Africa	525.8	24.53	568.7	20.80	0.66	575.2	20	1,204.7	20	2,353.8	22	5.85	6.32
Latin America	377.1	5.21	376.2	3.52	0.00	370.8	3	569.6	3	797.0	3	3.36	3.38

° Includes goat meat.

ᵇ Data for 1977 refer to trend values.

Source: Calculated from basic data in Food and Agriculture Organization of the United Nations. "Global Agricultural Programming System Supply Utilization Accounts Tape." Rome, June, 1980, and IFPRI Projections.

49

Table 4.5 Production, growth rate and percentage share in total meat of pigmeat, 1961/65 and 1973/77 averages.

('000 mt)

Region	1961/65	1973/77	Average Production Growth Rate 1961/65-1973/77 (percent)	% of Pig Meat in Total Meat 1961/65	% of Pig Meat in Total Meat 1973/77
Total 104 Countries	2,530.2	3,717.1	3.26	16.20	16.88
Asia	1,065.0	1,517.4	2.99	31.12	31.89
North Africa and Middle East	9.7	20.7	6.52	0.51	0.72
Sub-Saharan Africa	130.6	217.6	4.35	5.88	7.57
Latin America	1,324.9	1,961.4	3.32	16.43	17.04

Source: Food and Agriculture Organization of the United Nations: FAO Production Yearbook tapes 1975 and 1979; Rome, 1976 and 1980.

Table 4.6 Consumption, growth rate and percentage share in total meat of pigmeat, 1961/65 and 1973/77 averages with projections to years 1990 and 2000.

('000 mt)

Region	Average Consumption 1961/65	% of Total Meat	1973/77	% of Total Meat	Growth Rate 1961/65-1973/77 (percent)	1977[a]	% of Total Meat	1990	% of Total Meat	2000	% of Total Meat	Growth Rate 77-90 (percent)	Growth Rate 77-2000 (percent)
Total 104 Countries	2,588.5	17.33	3,835.8	17.69	3.33	4,049.7	18	6,964.9	16	10,306.9	14	4.26	4.15
Asia	1,132.9	32.20	1,661.7	33.33	3.24	1,729.5	33	3,064.6	30	4,563.4	27	4.50	4.31
North Africa/ Middle East	16.9	0.83	23.9	0.73	2.93	24.6	1	34.5	neg.[b]	44.1	neg.[b]	2.64	2.57
Sub-Saharan Africa	136.8	6.38	230.0	8.41	4.42	252.6	9	475.4	8	809.9	7	4.98	5.20
Latin America	1,301.9	17.99	1,920.2	17.97	3.29	2,042.9	18	3,390.5	17	4,889.5	17	3.97	3.87

[a] Data for 1977 refer to trend values.
[b] neg. = negligible.

Sources: Calculated from basic data in Food and Agriculture Organization of the United Nations, "Global Agricultural Programming System Supply Utilization Accounts Tapes." Rome, June, 1980, and IFPRI Projections.

Table 4.7 Production, growth rate and percentage share in total meat of poultry meat, 1961/65 and 1973/77 averages.

('000 mt)

Region	Average Production			% of Poultry Meat in Total Meat	
	1961/65	1973/77	Growth Rate 1961/65-1973/77 (percent)	1961/65	1973/77
Total 104 Countries	1,542.6	3,669.9	7.49	9.88	16.67
Asia	460.6	833.5	5.07	13.46	17.52
North Africa/ Middle East	235.9	622.4	8.42	12.34	21.64
Sub-Saharan Africa	222.1	402.5	5.08	10.00	14.00
Latin America	624.1	1,811.5	9.29	7.74	15.74

Source: Food and Agriculture Organization of the United Nations: FAO Production Yearbook tapes 1975 and 1979, Rome, 1976 and 1980.

Table 4.8 Consumption, growth rate and percentage share in total meat of poultry meat, 1961/65 and 1973/77 averages with projections to years 1990 and 2000.

('000 mt)

Region	Average Consumption					1977°	% of Total Meat	Consumption Projections					
	1961/65	% of Total Meat	1973/77	% of Total Meat	Growth Rate 1961/65-1973/77 (percent)			1990	% of Total Meat	2000	% of Total Meat	Growth Rate 77-90 (percent)	77-2000 (percent)
Total 104 Countries	1,579.8	10.58	3,871.2	17.85	7.75	4,495.4	19	9,362.3	21	16,325.7	23	5.81	5.77
Asia	474.3	13.48	887.1	17.79	5.36	984.4	19	2,130.3	21	3,873.2	22	6.12	6.14
North Africa/ Middle East	245.0	12.04	753.5	22.94	9.81	929.6	26	2,321.8	29	4,485.5	31	7.29	7.08
Sub-Saharan Africa	228.3	10.65	403.0	14.74	4.85	450.6	16	940.3	16	1,752.7	16	5.82	6.08
Latin America	632.2	8.74	1,827.5	17.11	9.25	2,130.8	19	3,969.8	19	6,214.4	21	4.90	4.76

° Data for 1977 refer to trend values.

Sources: Calculated from basic data in Food and Agriculture Organization of the United Nations, "Global Agricultural Programming System Supply Utilization Accounts Tape," Rome, June, 1980, and IFPRI Projections.

Table 5.1 Trends in domestic utilization of grains for feed* in developing countries, 1961/65 and 1973/77 averages

Region/Sub-Region	Grains for Feed		
	1961/65	1973/77	Growth Rate 1961/65-1973/77
	(in million metric tons)		(percent)
Total 104 Developing Countries	34.81	63.17	5.09
Asia	6.80	12.33	5.08
South Asia	3.76	5.68	3.49
East and Southeast Asia	3.04	6.65	6.73
North Africa/Middle East	10.24	16.12	3.86
Northern Africa	2.09	4.38	6.36
Western Asia	8.14	11.74	3.10
Sub-Saharan Africa	2.08	3.00	3.11
West Africa	0.92	1.19	2.12
Central Africa	0.18	0.27	3.54
Eastern and Southern Africa	0.98	1.54	3.87
Latin America	15.69	31.72	6.04
Central America and Caribbean	2.48	7.50	9.64
Upper South America	8.59	15.79	5.20
Lower South America	4.61	8.43	5.16

*Excluding rice bran.

Sources: Calculated from basic data in Food and Agriculture Organization of the United Nations "Global Agricultural Programming System Supply Utilization Accounts Tape" Rome, June 1980.

Table 5.2 Domestic utilization of grains for feed in developing countries, trend value 1977 and projections to 1990 and 2000

Region/Sub-Region	1977	1990	2000	Growth Rate	
				1977-1990	1977-2000
	(in million metric tons)			(percent)	(percent)
Total 104 Developing Countries	69.89	137.36	229.46	5.36	5.30
Asia	14.02	28.95	50.36	5.74	5.72
South Asia	6.18	10.74	15.89	4.34	4.19
East and Southeast Asia	7.84	18.22	34.46	6.70	6.65
North Africa/Middle East	17.36	40.25	73.93	6.68	6.50
Northern Africa	4.89	11.30	20.90	6.66	6.52
Western Asia	12.47	28.95	53.03	6.69	6.50
Sub-Saharan Africa	3.23	6.75	12.72	5.83	6.14
West Africa	1.26	3.06	6.69	7.05	7.51
Central Africa	0.29	0.51	0.79	4.43	4.42
Eastern and Southern Africa	1.68	3.17	5.24	5.03	5.08
Latin America	35.28	61.40	92.45	4.36	4.28
Central America and Caribbean	8.66	14.47	21.65	4.03	4.07
Upper South America	17.57	35.90	58.48	5.65	5.37
Lower South America	9.05	11.02	12.32	1.52	1.35

Sources: Calculated from basic data in Food and Agriculture Organization of the United Nations "Global Agricultural Programming System Supply Utilization Accounts Tape" Rome, July 1980.

Table 6.1   Trends in production of grains in developing countries,

1961/65 and 1973/77 averages

Region	Production of Grains		
	1961/65	1973/77	Growth Rate 1961/65-1973/77
	(in million metric tons)		(percent)
Total 104 Developing Countries	258.19	361.47	2.84
Asia	137.22	195.34	2.99
North Africa/ Middle East	41.48	54.26	2.26
Sub-Saharan Africa	29.40	35.45	1.57
Latin America	50.10	76.43	3.58

Note: Grains = wheat + rice (milled) + maize + millet + sorghum + other coarse grains.

Source: Food and Agricultural Organization of the United Nations FAO Production Yearbook Tapes 1975 and 1979, Rome 1976 and 1980.

Table 6.2 Production of grains in developing countries, trend value 1977 and projections to years 1990 and 2000

Production of Grains

Region/Sub-Region	1977	1990	2000	Growth Rate 1977-1990	1977-2000
	(in million metric tons)			(percent)	(percent)
Total 104 Developing Countries	383.25	575.91	802.65	3.18	3.27
Asia	206.60	310.82	430.72	3.19	3.25
South Asia	136.68	205.04	282.81	3.17	3.21
East and Southeast Asia	69.93	105.78	147.92	3.23	3.31
North Africa/Middle East	57.10	81.22	108.48	2.75	2.83
Northern Africa	17.69	24.45	31.80	2.52	2.58
Western Asia	39.41	56.77	76.68	2.85	2.94
Sub-Saharan Africa	37.35	50.83	68.74	2.40	2.69
West Africa	15.76	18.31	21.60	1.16	1.38
Central Africa	2.70	3.62	4.71	2.28	2.45
Eastern and Southern Africa	18.89	28.91	42.43	3.33	3.58
Latin America	82.20	133.04	194.71	3.77	3.82
Central America and the Caribbean	21.59	35.67	53.00	3.94	3.98
Upper South America	33.76	58.39	89.55	4.30	4.33
Lower South America	26.85	38.99	52.16	2.91	2.93

Note: Grains = wheat + rice (milled) + maize + sorghum + other coarse grains.

Source: Food and Agricultural Organization of the United Nations FAO Production Yearbook Tapes 1975 and 1979, Rome 1976 and 1980.

Table 6.3   Trends in domestic utilization of grains for food in developing countries, 1961/65 and 1973/77 averages

	(in millions metric tons)		
	Grains for Food		
Region	1961/65	1973/77	Growth Rate 1961/65-1973/77
			(percent)
Total 104 Developing Countries	196.2	274.3	2.83
Asia	122.2	168.4	2.71
North Africa/Middle East	22.6	39.5	3.35
Sub-Saharan Africa	23.0	31.1	2.55
Latin America	24.4	35.2	3.10

Sources: Calculated from basic data in ''Food and Agriculture Organization of the United Nations, ''Global Agricultural Programming System Supply Utilization Accounts Tape," Rome, June 1980.

Table 6.4   Domestic utilization of grains for food in developing countries, trend value of 1977 and projections to years 1990 and 2000

	(in million metric tons)				
	Grains for Food				
				Growth Rate	
Region/Sub-Region	1977	1990	2000	1977-1990	1977-2000
Total 104 Developing Countries	293.37	421.52	528.12	2.83	2.59
Asia	180.96	254.21	311.67	2.65	2.39
South Asia	116.85	166.01	209.07	2.74	2.56
East and Southeast Asia	64.11	88.20	102.60	2.48	2.07
North Africa/Middle East	42.28	60.33	72.71	2.77	2.38
Northern Africa	17.68	25.74	31.91	2.93	2.60
Western Asia	24.60	34.59	40.80	2.65	2.22
Sub-Saharan Africa	32.83	51.68	71.38	3.55	3.44
West Africa	14.48	23.60	33.89	3.83	3.77
Central Africa	3.23	4.89	6.48	3.23	3.07
Eastern and Southern Africa	15.12	23.20	31.01	3.35	3.17
Latin America	37.30	55.30	72.36	3.07	2.92
Central America and Caribbean	15.02	22.21	29.25	3.06	2.94
Upper South America	17.69	27.77	37.32	3.53	3.30
Lower South America	4.59	5.33	5.79	1.14	1.01

Sources: Calculated from basic data in Food and Agriculture Organization of the United Nations, ''Global Agricultural Programming System Supply Utilization Accounts Tape," Rome, June 1980 and IFPRI projections.

Table 6.5 Total domestic utilization of grains in developing countries, trend value 1977 and projections to years 1990 and 2000

Region/Sub-Region	Total Domestic Utilization of Grains				
	Trend	Projections		Growth Rate	
	1977	1990	2000	1977-1990	1977-2000
	(in million metric tons)			percent/year	
Total 104 Countries	415.49	639.26	869.12	3.37	3.26
Asia	219.94	319.84	410.74	2.92	2.75
South Asia	141.00	203.00	259.55	2.84	2.69
East and South East Asia	78.94	116.85	151.18	3.06	2.87
North Africa/Middle East	72.14	120.35	174.85	4.02	3.92
Northern Africa	26.01	42.31	60.04	3.81	3.70
Western Asia	46.13	78.04	114.81	4.13	4.04
Sub-Sahara Africa	41.68	67.35	97.07	3.76	3.74
West Africa	18.89	31.83	48.40	4.09	4.17
Central Africa	3.88	5.94	8.00	3.33	3.19
Eastern and Southern Africa	18.90	29.58	40.66	3.51	3.39
Latin America	81.74	131.70	186.46	3.74	3.65
Central America and Caribbean	26.43	41.04	57.06	3.44	3.40
Upper South America	39.60	71.51	107.74	4.65	4.45
Lower South America	15.71	19.15	21.66	1.54	1.41

Sources: Calculated from basic data in Food and Agriculture Organization of the United Nations "Global Agricultural Programming System Supply Utilization Accounts Tape." Rome, June 1980 and IFPRI projections.

Table 6.6 Trends In exports, imports and net trade of grains In developing countries, 1961/65 and 1973/77 averages

(in million metric tons)

Region	1961/65			1973/77		
	Imports	Exports	Net Trade	Imports	Exports	Net Trade
Total 104 Developing Countries	26.46	16.66	(9.80)	50.40	21.91	(28.49)
Asia	13.59	5.89	(7.70)	20.81	6.73	(14.08)
North Africa/ Middle East	4.92	1.11	(3.81)	11.60	0.73	(10.87)
Sub-Saharan Africa	1.91	0.66	(1.26)	4.45	0.60	(3.85)
Latin America	6.04	9.00	2.96	13.55	13.85	0.30

Note: Figures in brackets are net imports. Grain imports or exports include trade in bran and cake for feed use.

Sources: *Calculated from basic data in Food and Agriculture Organization of the United Nations, "Global Agricultural Programming System Supply Utilization Accounts Tape," Rome, June 1980.*

Table 6.7 Net surplus/deficit of grains In developing countries, trend value 1977 and projections to 1990 and 2000

(in million metric tons)

Region/Sub-Region	Surplus/Deficit of Grains		
	Trend	Projections	
	1977	1990	2000
Total 104 Countries	-32.24	-63.35	-66.47
Asia	-13.34	-9.02	19.98
South Asia	-4.32	2.04	23.26
East and Southeast Asia	-9.01	-11.07	-3.26
North Africa/Middle East	-15.04	-39.13	-66.37
Northern Africa	-8.32	-17.86	-28.24
Western Asia	-6.72	-21.27	-38.13
Sub-Saharan Africa	-4.33	-16.52	-28.33
West Africa	-3.13	-13.52	-26.80
Central Africa	-1.18	-2.32	3.29
Eastern and Southern Africa	-0.01	-0.67	1.77
Latin America	0.46	1.34	8.25
Central America and Caribbean	-4.84	-5.37	-4.06
Upper South America	-5.84	-13.12	-18.19
Lower South America	11.14	19.84	30.50

Note: Minus sign indicates deficit.

Sources: *Calculated from basic data in Food and Agriculture Organization of the United Nations, "Global Agricultural Programming System Supply Utilization Accounts tape," Rome, June 1980, and IFPRI projections.*

Appendix B

Developing Country Classification
Region/Sub-region/1977 GNP Per Capita Level @

Asia
South Asia

Less than $250: Bangladesh, Bhutan, India, Nepal, Pakistan, Sri Lanka (6)

East and Southeast Asia

Less than $250: Burma, Kampuchea, Laos, Vietnam (4)
$250 - 499: Indonesia, Philippines, Thailand (3)
$500 - 1,249: Korea, Democratic People's Republic; of Korea, Republic of; Malaysia, Mongolia, Papua New Guinea (5)
$1,250 and over: Fiji, Hong Kong, Singapore (3)

North Africa/Middle East
Northern Africa

$250 - 499: Egypt, Sudan (2)
$500 - 1,249: Algeria, Morocco, Tunisia (3)
$1,250 and above: Libya (1)

Western Asia

Less than $250: Afghanistan (1)
$250 - 499: Yemen, Arab Republic; Yemen, People's Democratic Republic (2)
$500 - 1,249: Jordan, Lebanon, Syria, Turkey (4)
$1,250 and above: Cyprus, Iran, Iraq, Kuwait, Oman, Saudi Arabia (6)

Sub-Saharan Africa
West Africa

Less than $250: Benin, Chad, Gambia, Guinea, Guinea-Bissau, Mali, Niger, Sierra Leone, Upper Volta (9)
$250 - 499: Ghana, Liberia, Mauritania, Nigeria, Senegal, Togo (6)
$500 - 1,249: Ivory Coast (1)

Central Africa:

Less than $250: Burundi, Central African Republic, Rwanda, Zaire (4)
$250 - 499: Angola, Cameroon (2)
$500 - 1,249: Congo (1)
$1,250 and above: Gabon (1)

Eastern and Southern Africa

Less than $250: Ethiopia, Lesotho, Madagascar, Malawi, Mozambique, Somalia, Tanzania (7)
$250 - 499: Botswana, Kenya, Uganda (3)
$500 - 1,249: Mauritius, Namibia, Swaziland, Zambia, Zimbabwe (5)
$1,250 and above: Reunion (1)

Note: In U.S. $ based on 1961-77 trend value of real GNP (1977 = 100)

Latin America
Central America and Caribbean

Less than $250: Haiti (1)
$250 - 499: Honduras (1)
$500 - 1,249: Cuba, Dominican Republic, El Salvador, Guatemala, Nicaragua (5)
$1,250 and above: Costa Rica, Jamaica, Mexico, Panama, Trinidad/ Tobago (5)

Upper South America

$250 - 499: Bolivia (1)
$500 - 1,249: Colombia, Ecuador, Guyana, Paraguay, Peru (5)
$1,250 and above: Brazil, Surinam, Venezuela (3)

Lower South America

$1,250 and above: Argentina, Chile, Uruguay (3)

Sources: World Bank, "Gross National Product 1960-78: Time Series Data by Country at Current and Constant Market Prices," Washington, D.C. 1979 (computer printout).

United Nations Department of International Economic and Social Affairs," World Population Trends and Prospects by Country, 1950-2000 (ST/ESA/SER R/33)

Appendix C
Data and Methodology
Data

The data on meat output, consumption and trade utilized in the study are taken from the international data base of the UN Food and Agriculture Organization (FAO). The annual data on meat output for the period 1961-65 are from the 1975 FAO Production Yearbook data tape, while those for the years 1966 to 1977 are from the 1978 data tape.[1] The consumption and trade data are from the Global Agricultural Programming Supply Utilization Accounts tape which contains the statistical base developed by the FAO for their more comprehensive global study, "Agriculture Toward 2000." These supply-utilization accounts show 1961-65 averages and annual estimates for the 1966-77 period. FAO also provided five-year-interval estimates of income elasticities by commodity for each country that were used for the consumption projections. Population estimates and projections are mostly from the UN Department of International Economic and Social Affairs, based on the 1978 assessment of world demographic data; for some countries, the study used the 1977 population estimates of the World Bank but the population projections were based on the U.N. projected growth rates. The annual data on Gross National Product (GNP) for each country which were used to provide estimates of trend growth rates of per capita income, at constant 1977 prices, are also from the World Bank.

Methodology: Past Trends

The analysis of past trends in the different variables used in this study are based on annual averages for the two five-year periods 1961-65 (early sixties) and 1973-77 (mid-seventies) for each individual country, sub-region and region. These averages were utilized for absolute measures of change and relative distributions. The annual growth rates worked out for this period represent the compound growth rates between the mid years of the two quinquennia.

Output Projections

For projecting the output of meat, the annual data for 1961-77 for each country were utilized and the trend growth rates as well as estimated trend values were obtained for meat, using the exponential trend equation (1) fitted to the time series data:

$$Y_t = ea + bt \qquad (1)$$

where $Y_t =$ estimate of variable in year t,

$a =$ constant term (the logarithm of the variable's estimate for the base year $t = o$),

---

[1]Annual data for 1961-65 were last published by the FAO in the 1970 Yearbook; later publications show only the averages for this period.

b = logarithm of the value one plus the annual rate of change $(1 + r)$ in the variable, and

t = period of years starting from the base year.

The projected outputs for 1990 and 2000 for each country were aggregated to give the results for sub-regions, regions, and the total for all the 104 study countries. A similar procedure was adopted to give the group aggregates by level and growth rate of per capita income and other typologies. Since country-level errors tend to be compensating for the aggregates, it can be expected that country measures are less reliable than those for country groups.

For calculating growth rates for country groups, the usual equation for measuring the average annual rate of change of the variables between two points of time was used.

$$Y_t = Y_0 (1 + r)^t \qquad (2)$$

where Y_0 = estimate of the variable for the base year, and

r = annual growth rate of the variable.

Thus, based on the assumption of continuing historical country trends in livestock production, the output projections are essentially extrapolations with the use of equation (1) over the period 1961-77. In some countries, however, where, after examination, the trends over 1961-77 are considered unsuitable, the 1970-77 period was utilized for projections. For developing countries with negative growth rates in output it is assumed that no further decline in production would occur and the trend estimate of output for 1977 was utilized both for 1990 and 2000. It may be added that although annual data on production by type of meat were available, only the total output was utilized in the projections because the rate of growth of the aggregate is more stable than those of the individual components.[1]

Consumption Projections: The demand for direct human consumption of meat was projected for each country using trend estimate of per capita consumption in 1977, per capita growth in income, income elasticity of demand and projected population. The historical pattern of utilization revealed by FAO data showed little or no allowances for wastage or for industrial usage (non-food manfactures) of meat. Hence these utilization categories were assumed to be zero.

Human Consumption

The trend estimates of the per capita use of each type of meat, namely, cattle and buffalo meat, mutton and goat meat, pig meat and poultry meat, for human consumption in 1977 were obtained for each country by using equation (1) over 1966-77, which is the period for which annual consumption data were available. Projections of per capita consumption in 1990 and 2000 were

[1]In some countries such a procedure may not be strictly advisable because each of these types constitutes a distinct enterprise, particularly where they do not compete for available resources.

obtained, making use of income elasticity coefficients and trend growth rates of per capita GNP (also based on 1966-77) in the equation (3).

$$C_{1990} = C_{1977}(1 + r_y xn)^{13 \text{ or } 23} \qquad (3)$$
$$\text{or}$$
$$2000$$

where c = per capita human consumption of the commodity during the indicated year,

r_y = trend annual growth rate of per capita real income, and

n = income elasticity of the demand for the commodity (when n is not constant, stepwise projections are made)

The total demand for direct human consumption for each country in 1990 or 2000 was calculated by multiplying the projected per capita consumption obtained from the above equation by the UN medium-variant population projection for 1990 and 2000. The estimates of projected consumption of each type of meat were added to obtain the projected consumption of total meat. In the case of per capita income growth, minimum and maximum limits on yearly growth rates were set at 0.5 percent and 6.0 percent respectively. Both production and consumption projections assume unchanged price relationships.

Alternative assumption regarding per capita-income growth

The estimated demand for meat in 1990 and 2000 was also calculated using per capita income growth which is 25 percent less than the trend growth. The basis for this assumption is explained in the Note in Appendix D.

Gross and Net Surplus and Deficit

The difference between the projected production and projected demand (total domestic utilization) gave the surplus or deficit for each country. For sub-regions, regions and other country groups, the surpluses and deficits of individual countries in the group were separately aggregated and denoted by gross surplus and gross deficit for the geographical area. The difference between the two represented the net surplus or net deficit, as the case may be, for the sub-region, region or country group.

Appendix D

A Note on the Population and Income Growth Assumptions in the Projections

Population Growth

The population of 104 developing countries included in the study, based on the UN medium variant population projections for individual countries, is estimated to increase from 2,092 million in 1977 to 2,911 million by 1990. This would represent an annual compound growth rate of about 2.6 percent or an average increase of 63 million people every year. Among the developing regions, the slowest growth of 2.4 percent is projected for Asia, whose expected population in 1990 of 1,635 million would account for 56 percent of the total population of the countries covered by the study. Half of the total annual increase in the Third World population would be from Asia. Population of North Africa/Middle East is expected to go up from 240 to 345 million between 1977 and 1990, or an annual growth rate of 2.9 percent. Sub-Saharan Africa would have the highest population growth rate of nearly 3.1 percent, with a projected population of 461 million. Latin America's population would increase by 10.5 million a year on an average, bringing it to the level of 470 million by 1990. Latin America and Sub-Saharan Africa would account for about 16 percent each of the Third World population in 1990. (See Table A).

By the end of the century, the total population of the study countries is expected to reach 3,640 million, showing an average compound growth rate of 2.4 percent per annum over 1977. The growth rates between 1990 and 2000 will be lower than those between 1977 and 1990 in all the regions, although in some countries, particularly in Sub-Saharan Africa, they may be still rising. The share of Sub-Saharan Africa in the Third World population would go up to about 17 percent in 2000.

Per Capita GNP Growth

If the 1966-77 growth rates in Gross National Product in the individual countries are extended to 1990, the resulting overall growth rate of GNP for all the study countries taken together over the period 1977-90 works out to 6.7 percent per annum. The fastest growth in GNP will be in North Africa/Middle East (7.5), followed by Sub-Saharan Africa (7.0). The slowest growth is projected for Asia per annum over the same period. As the overall annual growth in population during 1977-90 is projected at 2.6 percent, the resultant growth in per capita income of study countries taken together works out to about 4.0 percent. Here again the growth rate in per capita income in North Africa/Middle East will be the highest (4.5) and in Asia (3.3), the lowest. The growth rate in per capita income in Latin America (4.0) will be higher than that in Sub-Saharan Africa (3.9) despite the rapid increase of GNP in the latter.

Table A. Projected population, 1977, 1990 and 2000, and growth rates 1977-90 and 1977-2000

	Projected Population (in millions)			Growth Rates in Population (percent per year)	
	1977	1990	2000	1977-90	1977-2000
Asia	1,207	1,635	1,986	2.36	2.19
North Africa/ Middle East	240	345	441	2.85	2.69
Sub-Saharan Africa	311	461	614	3.06	3.00
Latin America	334	470	598	2.68	2.58
104 Study Countries	2,092	2,911	3,639	2.58	2.44

Source: UN Department of International Economic and Social Affairs." World Population Trends and Prospects by Country, 1950-2000," New York, 1979.

If the growth rates in GNP are further extended up to 2000, the overall per capita income growth over the 1977-2000 period works out to 4.3 percent per annum. The distribution of growth rates in per capita incomes among the regions over 1966-77, 1977-1990 and 1977-2000 is given in Table B.

The reference period for the growth in per capita income was chosen as 1966-77 to synchronize with the period over which the consumption was trended. Further, as explained in the methodology, maximum and minimum limits on annual per capita income growth rates were set at 6.0 and 0.5 percent respectively. In projecting the consumption, the per capita income growth rate for each country was utilized. However the average per capita income growth rates by region presented in Table B were derived from the compound growth rates of the trend estimate of GNP and the population estimate in the first and last years of the relevant period. Because of the weightage adopted, the average growth rates between 1977-90 and 1977-2000 show an acceleration compared to that over 1966-77.

Table B. Growth rates of per capita income 1966-77 and projected growth rates, 1977-90 and 1977-2000
Growth rates of per capita income (percent per year)

Region	1966-77	1977-90	1977-2000
Asia	3.30	3.30	3.76
North Africa/ Middle East	4.66	4.52	4.65
Sub-Saharan Africa	3.13	3.85	4.16
Latin America	3.51	4.05	4.24
104 Study Countries	3.70	4.03	4.30

Source: World Bank, "Gross National Product 1960-78. Time Series Data by Country at Current and Constant Market Prices," (Computer Printouts), Washington, D.C., 1979.

Further, the income growth rates in Sub-Saharan Africa may appear to be too high. These rates are influenced by the projected growth for Nigeria which was constrained at 6 percent per annum. Excluding Nigeria, the growth rates of per capita income in the rest of Sub-Saharan Africa work out to 1.6 and 1.7 percent per annum during 1977-90 and 1977-2000 respectively.

Even in the other regions, doubts are expressed whether the high growth rates in per capita income in the Third World countries will be repeated in the future. Due to improvements in agricultural technologies and institutions, increasing adoption of outward looking industrial policies and favorable external conditions, the 1966-77 period witnessed an extraordinarily high economic growth rate among developing countries. Subsequently, however, the two oil shocks of 1974 and 1979-80, the ensuing recession and protectionism in the OECD (Organization for Economic Cooperation and Development) countries and mounting debt repayment problems slowed economic growth in developing countries. At the present time, structural adjustments are being undertaken in several developing countries that show promise of improvements in the efficiency with which resources are allocated and used, enhancing the prospects of better economic performance in the future. Economic recovery and a retreat from protectionist policies in the OECD countries would reinforce the capacity of developing countries to raise their economic growth rate from the extraordinary low levels of 1980-83. Thus, while the assumption that the 1966-77 growth would continue to 1990 and 2000 may not appear realistic at the present time, it will be also unrealistic to assume that the future growth would be as slow as was experienced in the early 1980s. Moreover, the food gap analyses of IFPRI are done with particular concern for the food situation as it may develop at such time as the developing countries succeed in accelerating their growth from the 1980-83 levels. Also, in projecting the demand, estimates are also represented of the likely demand under an alternative lower income growth assumption, namely, per capita income growth, say, 25 percent less than the 1966-77 trend growth.

Bibliography

Food and Agriculture Organization of the United Nations. "Global Agriculture Programming System Supply Utilization Accounts Tape," Rome, June 1980.

"Parameters of the Demand Functions," fifth run, April 1978 (computer printout).

FAO Production Yearbook, 1982, Rome: FAO, 1983.

"FAO Production Yearbook Tape, 1975," Rome, 1976.

"FAO Production Yearbook Tape, 1979," Rome, 1980.

Paulino, Leonardo A. "Food in the Third World: Past Trends and Projections to 2000," International Food Policy Research Institute, Washington, D.C., 1984 (mimeographed).

Sarma, J.S. and Yeung, Patrick. "Livestock Products in the Third World: Past Trends and Projections to 1990 and 2000," International Food Policy Research Institute, Research Report 49, Washington, D.C., 1985.

United Nations, Department of International Economic and Social Affairs. *World Population Trends and Prospects by Country, 1950-2000* (ST/ESA/SER.R/33), 1979.

World Bank. "Gross National Product, 1960-78: Time Series Data by Country at Current and Constant Market Prices," Washington, D.C., 1979 (computer printout).

1979 World Bank Atlas, Washington, D.C.: World Bank, 1979.

Chapter 3

The Introduction of Isolated Soy Protein Food Ingredients in Sweden: Prospective Impacts on Trade, Food Policy, and Agricultural Resource Use

*Matthew G. Smith**

*Matthew G. Smith was a research specialist at the University of Minnesota, St. Paul, working under the direction of G. Edward Sihoh at the time this paper was written.

Abstract

Sweden has been a net exporter of meat products for almost two decades. Incorporation of isolated soy protein ingredients in processed meat products offers an opportunity to raise export earnings, to lower domestic retail prices, and to lessen pressure on livestock production in the event of shortages or increases in consumer demand.

Production and marketing in Sweden are governed by a unique system of price supports and subsidies. These reflect a wide range of social, economic, environmental and national security interests.

For instance, safeguarding the country's food supply from interruption became an important component of Sweden's defense policy following severe shortages during World War I. Maintaining self-sufficiency in basic foodstuffs is a primary policy goal. These government policies are expected to continue. They have protected high domestic meat prices in order to achieve production and farm income goals.

Measures, such as incorporation of isolated soy protein ingredients in processed meats, which would reduce the cost of meat products to consumers while maintaining producer income at desired levels, also could reduce budgetary pressures on the government.

Naturally, government policies affect retail and export market activity and product distribution. Today, approximately 40 percent of Swedish domestic meat consumption consists of cured, processed and canned meats.

Beef carcasses and pork cuts make up about 90 percent of Sweden's total meat exports. Processed products, such as sausages, other meat preparations, and other products now account for a very small share of total meat exports—between 1 and 2 percent in recent years.

Increases in meat exports could offer significant benefits to the Swedish economy. While very small relative to total exports, the meat sector is relatively large in comparison to the country's trade balance and balance of payments in recent years.

This analysis indicates that using isolated soy protein in processed meats (sausages) in Sweden's meat industry would produce a 31 percent increase in final product yield from a given meat input, and a retail price reduction of about 7 percent.

Addition of soy protein isolates to processed ham would produce a 5 percent extension of green ham input, lower ingredient costs, and reduce retail prices about 3.2 percent.

Processed meat products could become a significant contributor to Sweden's export earnings while also increasing the supply of meat products available for domestic consumption.

73

By increasing the output of processed meat products and trimming ingredient costs, isolated soy proteins present Sweden with an opportunity to increase export earnings from processed meat products through reducing the differential between domestic and international prices.

Reduced ingredient costs for reformulated products also would bring Swedish prices more in line with those in the world market, thus reducing or conceivably even eliminating the need for export subsidies on these products.

The total trade gain to Sweden from use of isolated soy protein ingredients is projected at about 576 million kronor annually.

As discussed elsewhere in this volume, use of isolated soy protein ingredients, by reducing the proportion of animal products in processed meat production, also offers savings in all inputs required for meat production.

At the current level of carcass utilization in processed products, introduction of soy protein isolates in Sweden's processed meat industry would save nearly 41,000 tons of pork annually, or the equivalent of 164,000 tons of grain.

While soy protein isolate would have to be imported, the relative quantity required to substitute for the supplies of fuel and fertilizer needed to support a given level of meat product production and consumption is quite low.

Introduction

Until recently, vegetable protein to extend or replace animal protein has been most widely available to food processors in the form of soy flour or textured soy protein. Both of these ingredients are obtained by changing the physical form of soy flakes, a product containing 50 percent protein. Reformulated meat products containing these ingredients, as well as meat analogs made from vegetable protein, have not met with wide consumer acceptance. Neither of these groups of products fit within the quality standards and cost structures of traditional food cultures.

Advances in processing have led to more efficient procedures for extracting or "isolating" the protein from the soy flake, resulting in the availability of isolated soy protein food ingredients containing 90 percent Protein. These isolates may be produced in powdered, granular, or fibrous forms. As an ingredient in reformulated meat products, the incorporation of up to 6 percent isolated soy protein has proven successful in maintaining the sensory and nutritional qualities of traditional meat products.

The incorporation of vegetable protein as a partial substitute for animal protein in meat products offers the opportunity to produce greater quantities of meat products from a given base of livestock production. Because of losses in the transformation of vegetable to animal protein in meat production, the use of isolated soy protein also allows potentially large savings in the total resources required for the production of meat products. Improved efficiency of meat products output also offers potential advantages to consumers in the form of reduced ingredient costs for the reformulated meat product.

This paper is an examination of the prospective economic impacts of the introduction of isolated soy protein food ingredients in Sweden. Part I provides an overview of recent trends in meat production, consumption, and prices; assesses the feedgrain resources currently employed in meat production; and outlines some current issues in food and agricultural policy that could be affected by the introduction of isolated soy protein. Part II projects soy protein's impacts on prices of processed meat and ham products. Part III examines the potential trade impacts of soy protein ingredients. Part IV evaluates the prospective impact of soy protein on agricultural resource use under varying levels of consumption of processed meat products.

Sweden is a particularly interesting case to consider since it has been a net exporter of meat products for approximately two decades. Hence, an analysis of the data from this country will provide a means of examining the potential of isolated soy protein food ingredients as a means to increase export earnings. Domestically, soy protein also offers opportunities to reduce retail meat product prices and to lessen pressure on critical inputs for livestock production in the event of shortages or increases in consumer demand.

The Swedish Meat Sector: Recent Trends and Outlook for the Future
Meat Production, Consumption, and Trade

Production of meat incraased by nearly one quarter in Sweden between 1970 and 1980, and by 1981 reached a total of 528,000 metric tons (Table 1).

Table 1

Meat Production in Sweden
(Carcase weight in metric tons)

Year	Beef[1]	Veal[1]	Horsemeat[1]	Mutton[1] and Goatmeat	Pork[1]	Poultry[2]	Total, all meats
1982	150,537	10,312	2,649	5,268	325,020		493,786[3]
1981	146,517	11,755	2,600	4,910	320,540	42,300	528,622
1980	145,653	11,688	2,594	4,801	317,371	43,900	526,007
1979	140,221	11,395	2,721	4,788	313,650	41,700	514,475
1978	135,338	11,807	2,806	4,831	306,760	40,400	501,942
1977	135,535	12,613	2,786	4,959	303,571	37,800	497,264
1976	134,313	14,697	2,523	4,722	293,690	40,700	490,645
1975	128,438	15,108	2,046	4,280	282,716	38,910	471,498
1974	128,791	15,057	1,811	4,232	276,970	39,060	465,921
1973	113,473	13,273	1,976	3,397	263,746	34,700	430,565
1972	112,247	13,124	2,765	3,087	276,025	31,430	438,678
1971	127,308	15,494	2,926	3,216	256,306	27,320	432,570
1970	144,207	16,829	3,718	3,444	236,082	26,750	431,030
1965	132,512	25,556	8,551	2,040	215,554		377,434
1960	102,566	18,749	7,932	1,256	213,792		351,103

[1] Source: Sweden. Statens jordbruksnamnd. *Jordbruksekonomiska meddelanden*, various issues.

[2] Source: FAO, International Computer System, printout on Supply Utilization Accounts provided by International Food Policy Research Institute.

[3] Excludes poultry.

Of this amount, pork accounted for over 320,000 tons (61 percent of the total). The remainder of meat output is divided among poultry (42,000 tons), mutton and goatmeat (4,900 tons), and horsemeat (2,600). Pork production has increased at an average rate of approximately 2 percent annually since 1975. Total output of beef and veal in 1982 was slightly above the 1965 level, but the production of veal only has declined by more than half since that time. Beef output increased until 1970, and then fell to a low of 113,000 tons in 1973. Since that time, beef production has increased dramatically, to over 150,000 tons in 1982. Production of poultry meat has increased most rapidly, rising at an average annual rate of over 5 percent during the 1970's. Mutton and goatmeat production has also increased significantly in recent years, but remains a small fraction of total meat output. Horsemeat production has been stable since the mid-1970's, at a level two-thirds below that of the early 1960's.

In contrast to meat production, consumption in Sweden has declined each year since its peak in 1979.

Consumption rose dramatically between 1972 and 1976, spurred by a retail price freeze on meat products from 1973 to 1975 and government subsidies to help insulate consumers from producer price increases after the freeze was lifted. Total meat consumption increased by nearly 100,000 tons in 4 years, for an average rate of increase of over 5 percent annually. As a result of a general economic slowdown and sharp reductions and then elimination of subsidies (prompted by efforts to reduce government expenditures), total meat consumption fell approximately 3 percent in both 1981 and 1982. Despite the recent downturn, total meat consumption in Sweden remains nearly one-third higher than the 1960 level.

Consumption of various meat types in Sweden is distributed in similar proportions to meat output. Pork, beef and veal, and poultry meat combined account for almost 95 percent of the total meat consumed. Pork consumption, which reached 294,000 tons in 1979, fell to 273,000 tons in 1982. This is still 24 percent higher than in 1970 and 50 percent higher than in 1960. Beef consumption reached a high of 151,000 tons in 1973 and by 1982 stood at 131,000 tons, approximately equal to the 1970 level and only 6 percent higher than in 1965. The 1982 poultry meat consumption of 41,000 tons represents a doubling since 1965 and more than a tripling since 1960. Horsemeat and goatmeat and mutton are consumed in relatively small quantities, and horsemeat consumption declined steadily throughout the 1960's and 1970's.

The growing surpluses of domestic meat production over consumption have led to increasing meat exports from Sweden.

The country has been a consistent net exporter of meat during the past two decades, but rapidly increasing domestic consumption during the mid-1970's resulted in a decline from the previous high of 51,000 tons in 1972 to approximately 2,000 tons in 1975 and 1976. Sluggish consumption since that time, combined with continued increases in production, pushed net exports of all meats to a new high of 78,000 tons in 1982.

Beef and pork are the principal meats traded, together accounting for nearly all of Sweden's net trade in meat. Pork exports have ranged between 20 and 50 thousand tons in recent years, and have been increasing in tons since 1975 and 1976, but regained its position as a net beef exporter in 1979. 1982 net beef exports totalled nearly 27,000 tons. Sweden is largely in balance in other products, exporting small amounts of poultry meat and importing small quantities of mutton and horsemeat (less than 1,000 tons of each type in recent years).

Table 2

Meat Consumption in Sweden
(Carcase weight in metric tons)

Year	Beef[1]	Veal[1]	Horsemeat[1]	Mutton[1] and Goatmeat	Pork[1]	Poultry[2]	Total, all meats
1982	131,276	9,839	2,729	5,300	272,733	40,900[3]	462,777
1981	133,249	11,647	2,838	5,831	282,432	42,339	478,336
1980	140,085	11,364	3,527	5,427	288,817	42,981	492,201
1979	140,309	11,930	3,687	5,645	293,777	41,219	496,567
1978	137,160	12,223	3,900	5,648	280,736	40,449	479,846
1977	141,025	13,665	3,822	6,269	280,450	35,153	480,384
1976	151,443	14,952	4,388	5,175	275,870	39,909	491,739
1975	145,108	15,349	5,202	4,891	257,466	38,188	466,204
1974	132,968	15,018	5,604	4,653	246,540	37,135	441,918
1973	115,936	12,695	6,570	4,638	222,900	34,144	396,973
1972	110,325	12,841	5,757	4,326	227,737	31,698	392,684
1971	118,777	16,013	5,504	4,276	226,715	28,335	339,620
1970	131,994	17,312	7,003	4,434	220,305	27,124	408,172
1965	123,833	21,217	11,013	2,227	199,477	19,800[3]	377,567
1960	111,770	28,128	14,350	1,940	182,189	11,500[3]	349,877

[1] Source: Sweden. Statens jordbruksnamnd. *Jordbruksekonomiska meddelanden*, various issues.

[2] Source: FAO, International Computer System, printout on Supply Utilization Accounts provided by International Food Policy Research Institute (all years except 1960, 1965, 1982)

[3] Source: Sweden. Statistiska central byron. *Jordbruksstatistisk arsbok*, various issues.

Table 3

Sweden: Net Foreign Trade in Meat
(Carcase weight in metric tons)
(Numbers in parentheses indicate net imports)

Year	Beef[1]	Pork[1]	Total, All Meats[2],[3]
1982	26,550	51,727	78,336[4]
1981	7,787	35,745	44,315
1980	2,520	29,015	30,523
1979	3,933	20,036	22,515
1978	(5,172)	27,122	19,508
1977	(4,830)	22,581	17,395
1976	(17,142)	19,110	1,793
1975	(17,439)	23,930	1,930
1974	(6,443)	32,596	21,426
1973	(3,873)	43,121	34,110
1972	2,132	52,318	51,021
1971	10,167	23,638	30,209
1970	15,907	11,161	22,239
1965	9,405	14,002	19,429[4]
1960	(4,893)	23,184	22,023[4]

[1] Source: Sweden. Statens jordbruksnamnd. *Jordbruksekonomiska meddelanden,* various issues.

[2] Includes veal, mutton and goatmeat, horsemeat, and poultry.

[3] Source for poultry data: *FAO Trade Yearbook,* various issues.

[4] Excludes poultry.

Producer and Consumer Meat Prices

An explanation of the recent divergence between output and consumption in the Swedish meat sector is evident from data on average meat prices at the producer, wholesale, and retail levels. Under the Swedish system of price regulations for food products, delegations representing consumers, producers, and the government agree on semi-annual adjustments to aggregate farm income, which are then distributed over various commodity groups to obtain changes in product prices. Wholesale meat prices are set above world market prices, with a system of variable import levies designed to keep Swedish prices within upper and lower price limits.

Meat prices are targeted at the wholesale level rather than the producer level in order to calculate import levies and because of the widespread membership of Swedish farmers in meat marketing and processing cooperatives. Farm incomes are influenced by cost and profit levels in the cooperative food industry (which handles a substantial majority of Swedish agricultural output) as well as farm production costs and revenues, and prices at the producer level thus represent a residual after the wholesale Price has been accounted for. Since 1973, a program of meat price subsidies paid by the government to processors for distribution to producers in the form of price supplements had allowed wholesale meat prices to be insulated from the full effects of producer price increases. The subsidy program was eliminated in December 1983. Retail meat prices are linked to wholesale prices by processing and marketing margins, and the negotiated wholesale price is thus a fulcrum from which consumer prices are derived as well.

Meat price subsidies, whether technically considered subsidies to producers or consumers, have the effect of partially uncoupling changes in producer prices from changes in retail prices. This effect goes far in explaining developments in Swedish meat production and consumption since the early 1970's. Table 4 presents average annual prices of beef, pork, and broilers at various levels since 1968. The effects of the 1973-75 retail price freeze for beef and pork are quite apparent. Wholesale and retail Prices decreased slightly from 1972 to 1975, while average producer prices (the wholesale-derived price supplemented by government funds) increased by nearly 60 percent for beef and by more than 50 percent for pork. As a result, consumption of beef and pork increased dramatically, even outpacing increases in production. Pork surpluses were reduced and Sweden became a net importer of beef.

After the expiration of the retail price freeze at the end of 1975, subsidies continued in use in order partially to insulate consumers from the effects of producer price increases. Thus while retail meat prices rose, they did not fully reflect the high prices paid to Swedish livestock producers, as they had during the pre-1973 period. Retail beef and pork prices increased 44 and 60 percent, respectively, between 1975 and 1980. This slowed and eventually reversed the growth in meat consumption, while steady increases in producer prices led to continued growth in output and surpluses.

Table 4
Producer, Wholesale, and Retail Meat Prices in Sweden
(Swedish kronor per kilogram; carcase weight in the case of producer and wholesale prices)

Year	Beef			Pork			Poultry	
	producer[1]	wholesale[2]	retail[3] (gr. beef)	producer[1]	producer[2]	retail[3] (gr. pork)	wholesale[2]	retail[3] (frozen broilers)
1983[4]	19.74	20.24		11.78	16.22		12.59	
1982	19.08	18.74		11.14	14.92		17.23	
1981	17.64	16.04		10.24	12.00		11.22	
1980	16.45	14.36	32.57	9.08	9.83	23.58	10.28	17.13
1979	15.39	13.19	29.65	8.15	8.76	21.00	8.97	14.94
1978	14.69	12.67	28.95	7.64	7.90	20.63	9.19	14.94
1977	14.26	11.56	26.20	7.41	7.10	19.80	8.42	13.09
1976	13.16	10.10	22.80	6.92	6.63	17.34	8.11	12.26
1975	11.35	8.96	20.37	6.32	6.13	16.38	7.72	11.43
1974	9.88	8.65	19.56	5.94	6.30	16.76	7.32	10.99
1973	9.15	9.24	20.75	5.51	6.86	17.69	6.51	10.39
1972	8.29	9.21	20.19	4.56	6.23	16.48	6.01	9.63
1971	7.39	7.95	17.32	4.63	5.70	15.13		8.97
1970	6.87	7.57	16.00	5.01	6.26	14.65		8.36
1969	6.74	7.21	15.25	4.72	5.84			7.89
1968	6.76	6.97	14.59	4.31	5.27			8.25

[1]Source: Sweden. Statistiska centralbyron. *Jordbruksstatistisk arsbok,* various issues.
[2]Source: Sweden. Statens jordbruksnamnd. *Jordbruksekonomiska meddelanden,* various issues.
[3]Source: Sweden. *Konsumentpriser och indexberakningar,* 1981.
[4]Producer prices for first 8 months of year, wholesale for first 10 months.

Developments since 1980 suggest even steeper increases in retail meat prices. Budgetary stringency resulting eventually in the elimination of subsidies has left consumers exposed to the effects of high producer prices. In 1982 alone, wholesale prices rose 24 percent for pork and 17 percent for beef.

Meat consumption patterns have also been affected by changing relative prices of beef, pork, and poultry meat. The over 250 percent increase in consumption of poultry meat since 1960 reflects a near-halving of the retail price of frozen broilers relative to ground beef and a more than one-third reduction in the price of broilers relative to pork chops. The 1973-75 price freeze, which did not include poultry meat, interrupted the trend of declining relative poultry prices and probably reduced the growth in consumption during the mid-1970's from what it otherwise would have been. Poultry was included in the meat subsidy program beginning in 1976, but the allocations were withdrawn after 1981.

Feedgrain Inputs in the Meat Sector: Domestic Production and Foreign Trade

Table 5 presents data of feeding patterns and feed conversion in Swedish meat production. Pork rations consist almost entirely of grain supplemented with protein and mineral concentrates. Barley and oats are the major feedgrains used in pork production. Poultry are fed a larger proportion of soy meal concentrates, and use corn (maize) in addition to barley and other small grains, including wheat and rye. Cattle fed more intensively receive large amounts of barley and oats in addition to forages, while in less intensive feeding programs, the relative proportions of grains and forages are reversed.

Production of major feedgrains in Sweden since 1960 is summarized in Table 6. The data indicate the growth in output of the main feedgrains, in particular barley, which has increased approximately 200 percent since 1960. Oats production has also increased since the 1960's, although output remained relatively stable during the 1970's.

Sweden's net foreign trade in major grains is presented in Table 7. In recent years Sweden has exported 100,000 to 300,000 tons of barley and 200,000 to 300,000 tons of oats annually. The country also exports large amounts of wheat (anywhere from 250,000 to 800,000 tons in each of the past several years) and is also a net exporter of rye. Sweden relies on imports for two major ingredients in its livestock and poultry feeds. Corn imports have ranged between 50,000 and 85,000 tons in most recent years, and imports of soybean meal have averaged approximately 220,000 tons annually since 1975.

The Processed Meat Industry in Sweden

Nearly all Swedish farmers are affiliated with the country's major farm cooperative, the Federation of Swedish Farmers. More than 80 percent of all Swedish farm produce passes through the agricultural cooperatives on its way to the consumer. Producer cooperatives own about 46 percent of Sweden's

Table 5

**Feed Conversion and Major Feedstuffs
Used in Swedish Livestock Production [1]**

	Pork	Poultry	Beef (more intensive)	Beef (less intensive)
Kg. feed per kg. carcase weight	4.6[2]	2.6	9.2	13.2
Carcase wt. per animal (kg.)	76	1.0	220	275
Feed content (%):				
Grain	80.3	65	67.8	28.6
Peas	0.5			
Milling waste	4.8			
Soy - meal	4.3	22.4	5.2	2.3
Fish - meal	4.4	4		
Meat - meal	1.0	2		
Milk - food			1	0.5
Hay/silage/pasture			13.8	61.8
Straw			9.9	5.5
Slaughterhouse Fat		3		
Minerals	4.6	3.6	2.3	1.2

[1] Data provided by P. Andersson, Swedish University of Agricultural Sciences, Research Information Center, private communication.

[2] 100 Kg. feed up to 25 kg. live weight, boar and sow feed included. 250 kg. from 25 to 105 kg. live weight.

Table 6

Grain Production in Sweden[1]
(1,000 metric tons)

Year	Wheat (spring & winter)	Barley	Rye	Oats
1982	1,490.4	2,378.2	209.0	1,662.8
1981	1,066.3	2,451.7	177.3	1,815.6
1980	1,193.3	2,171.9	222.9	1,566.7
1979	1,003.0	2,345.3	192.1	1,524.1
1978	1,290.1	2,434.4	293.6	1,550.4
1977	1,522.4	1,966.1	335.0	1,416.3
1976	1,764.6	1,825.6	424.1	1,251.3
1975	1,454.6	1,902.8	322.2	1,320.9
1974	1,793.1	2,356.1	429.4	1,656.3
1973	1,335.5	1,767.5	321.5	1,209.4
1972	1,150.4	1,883.2	363.4	1,629.5
1971	994.9	2,029.0	301.5	1,866.7
1970	962.1	1,903.7	225.3	1,685.8
1965	1,070.4	1,429.5	174.7	1,338.9
1960	986.0	801.8	240.5	1,247.8

[1]Source: Sweden. Statistiska centralbyron. *Jordbruksstatistisk arsbok* (Stockholm, various issues).

Table 7

Sweden. Net Foreign Trade in Feedstuffs
(1,000 metric tons)
(Numbers in parentheses indicate net imports)

Year	Wheat[1]	Barley[1]	Rye[1]	Oats[1]	Corn[1] (maize)	Soybean[1] Meal
1982	394.1	112.2	23.6	330.9	(83.4)	
1981	255.6	107.2	1.8	274.5	(71.7)	(221.6)
1980	279.0	233.5	44.3	282.2	(84.0)	(207.6)
1979	310.2	295.8	78.0	290.4	(63.5)	(232.7)
1978	510.8	113.1	97.7	185.0	(74.0)	(237.7)
1977	783.7	80.1	123.6	(12.9)	(51.1)	(215.3)
1976	788.4	141.1	125.0	49.3	(73.0)	(212.6)
1975	786.1	310.5	116.0	172.2	(87.0)	(212.1)
1974	665.8	(11.2)	108.7	(41.1)	(131.9)	(243.0)
1973	352.3	86.0	100.1	125.2	(58.0)	(190.3)
1972	265.5	291.0	64.2	542.7	(39.7)	(197.2)
1971	124.4	437.6	103.0	401.6	(43.0)	(166.9)
1970	239.5	91.2	27.5	69.8	(41.0)	(168.0)
1965	284.0	114.4	(58.5)	50.6	(48.7)	(168.4)
1960	29.3	(67.9)	(35.8)	8.1	(47.9)	

[1] Source: Sweden. Statens jordbruksnamnd. *Jordbruksekonomiska weddelanden*, various issues.

[2] Source: *FAO Trade Yearbook*, various issues.

food processing industry, consumer cooperatives another 12 percent, and the government about 5 percent. Private Swedish and foreign companies control the remainder of the industry.

Within the meat sector, the Swedish Farmers' Meat Marketing Association accounts for over 80 percent of the total slaughter. The remainder of the slaughter is shared by consumer cooperatives and private concerns. Meat packing and distribution is shared more equally, with producer cooperatives and private packers each holding nearly 40 percent of the market. The producers' cooperative is responsible for approximately 38 percent of Sweden's total cured meat production. The Swedish Farmers' Meat Marketing Association is also the country's dominant meat exporter, handling 75 percent of total beef exports and nearly half the pork trade in 1982.

Table 8 presents data on the consumption of fresh and processed meat in Sweden in 1960 and the years 1981-83. As the figures indicate, cured, processed, and canned meats currently represent approximately 40 percent of swedish meat consumption, down from nearly 50 percent in 1960. Mixed cured meats, which includes sausages, are the most common processed meat products. Consumption of processed products has fallen more rapidly than consumption of fresh cuts during the 1980's. The most dramatic decline has come in mixed cured meats, for which domestic use fell off 14,000 tons in two years.

Table 8
Consumption of Fresh Cuts and Processed Meat in Sweden, 1960 and 1981-82

Year	1960	1981	1982	1983
	(Thousand metric tons)			
Fresh & frozen meat	171.5	243.9	238.9	239.8
Poultry meat, fresh & frozen	11.5	41.3	40.6	40.5
Cured meats & provisions, unmixed (incl. hams)	16.0	28.8	27.7	26.0
Cured meats & provisions, mixed (incl. sausages)	124.6	145.6	140.6	131.7
Canned meat	20.2	19.9	18.3	17.1
Canned meat soups	4.3	11.1	10.4	9.2

Source: *Jordbruksekonomiska meddelanden, 1983 (11): 32.*

Table 9 details average retail prices of processed meat products compared with the average retail prices of all meat products, fresh cuts and processed, for the period 1981-83. The data suggests that the greater processed, for the period 1981-83. The data suggests that the greater decline in consumption of processed products was due to larger retail price increases for those products than for pork as a whole.

Table 9

Average Retail Prices, Average of All Pork Cuts
Versus Processed Meat Products, 1981-83

Year (December average)	1981	1982	1983	Percent Change, 1981-83
	— kronor per kilogram —			
All pork (average)	35.40	40.52	47.01	33
Unmixed cured meats	62.22	72.72	86.98	40
Mixed cured meats	26.90	31.31	38.06	41

Source: Data provided by Han-Erick Uhlin, Swedish University of Agricultural Sciences.

Sweden's exports of meat carcases, fresh cuts, and processed meat products during 1981-83 are summarized in Table 10. As the table indicates, beef carcases and pork cuts make up approximately 90 percent of total meat exports. Processed products, such as sausages, other meat preparations, and other products, account for a very small share of total Swedish meat exports -between 1 and 2 percent in recent years. Processed meats are exported to a variety of countries. The Federal Republic of Germany is the single largest customer, with most of the remainder going to other European nations. A very small fraction of processed meat exports is destined for Asian and African countries.

Table 10

Sweden: Exports of Various Meat Products

1981-83

Year	1981		1982		1983	
	tons	percent of total meat exports	tons	percent of total meat exports	tons	percent of total meat exports
Beef carcases	10,820	22.4	25,309	31.4	12,791	23.7
Veal cuts	248	.5	631	.8	229	.4
Beef cuts	1,124	2.3	5,298	6.6	4,053	7.5
Hog carcases	190	.4	1,120	1.4	87	.2
Pork cuts	34,819	72.2	46,962	58.3	35,653	66.0
Sausages and similar products	146	.3	223	.3	92	.2
Other meat preparations	251	.5	411	.5	101	.2
Other meat, parts and products of meat or parts	598	1.2	634	.8	983	1.8

Source: Sweden. Central Statistical Bureau, Foreign Trade Statistics, various issues.

The Meat and Feedgrains Sector in the Context of Sweden's Total Trade and Balance of Payments

The balance of payments situation since 1974 is summarized in Table 11. Sweden has alternated between surpluses and deficits overall in recent years, but has experienced consistent deficits in its current accounts balance. This has largely been offset by government borrowing and other inflows of short term capital from abroad. Within the current accounts, the net trade balance slipped into deficit in 1979 and 1980 before posting a small surplus in 1981.

As Table 12 indicates, the meat sector, while very small in terms of total exports, is relatively large in comparison to Sweden's trade balance and balance of payments in recent years. Thus improvements in the trade performance of this sector of the economy could offer significant benefits.

The Swedish krona's foreign exchange rate is pegged to a composite index of the currencies of the country's major trading partners. The average annual exchange rate of the krona versus the U.S. dollar from 1970 to 1983 is presented in Table 13. From 1972 to 1980, the exchange rate ranged between 4.1 to 4.7 kronor per dollar. Since that time the krona's value relative to the dollar has slipped considerably. By 1983 the krona had slipped to about 7.7 per dollar.

Current and Emerging Issues in Swedish Food and Agricultural Policy

Swedish food and agricultural policy reflects a wide range of social, economic, environmental, and national security interests. Although the nation has not been at war for over 150 years, Sweden suffered severe food shortages during the First World War due to the disruption of shipping. With this experience in mind, the notion of safeguarding the country's food supply from interruption became an important component in the Swedish strategy of "total defense." Since World War II, this attitude has found explicit expression in the policy goal of maintaining a high degree of self-sufficiency in basic foodstuffs. Although the precise definition of self-sufficiency and the extent to which it should be pursued have been the source of some discussion, the overall thrust of the policy has remained quite consistent throughout the postwar period.

A number of other goals have also been important in shaping Swedish food and agricultural policy. Concern for the maintenance of farm income at levels equivalent to those of other comparable groups in society has influenced negotiated prices for agricultural commodities and has led to special support programs for small producers. The aesthetic value of a cultivated, orderly countryside has added to the impetus to keep arable land in production. Regional development goals have led to special support Programs to allow efficient farms to remain in operation in remote areas of the country, such as Northern Sweden.

As a result of these influences, Swedish policies have protected domestic meat prices, and to a lesser extent grain prices, at high levels in order to achieve production and farm income goals. Some indication of the extent of this

Table II

Sweden. Balance of Payments, 1974-81[1]

Millions of kronor

	1974	1975	1976	1977	1978	1979	1980	1981
Balance on:								
Current Account	-2,450.6	-1,451.9	-71,764.4	-9,810.8	-1,159.7	-10,307.9	-18,969.9	-14,819.0
thereof: trade balance	1,740.5	4,436.4	709.1	1,198.2	11,648.0	-2,930.1	-9,451.9	543.3
Direct investment and other long-term capital	1,879.3	5,626.2	1,609.3	4,138.8	-3,145.3	-3,556.6	-1,855.2	-5,075.0
Other short-term capital	-1,420.2	1,285.6	2,615.1	-476.1	3,733.7	1,966.3	3,396.5	7,355.8
Net errors and omissions	-1,356.1	-4,739.9	648.7	1,334.3	-2,364.7	-592.7	-5,703.1	-2,901.7
Counterpart items and exceptional financing (incl. loans to official sector and public sector bonds issued abroad)	-272.3	1,114.1	105.6	9,507.3	4,893.4	9,466.0	23,126.1	17,846.1
Net balance of payments	-3,614.5	6,100.1	-2,192.6	4,693.5	1,951.7	-2,819.3	—	2,406.2

Source: IMF, Balance of Payments Statistics, Vol. 33, Part 1, 1982. Numbers may not add to totals due to rounding.

Table 12

Sweden. Exports of Meat Products Relative to Total Exports, 1981[1]

	Million Kronor	Percent of total exports
Beef (fresh, chilled, or frozen)	128.5	.09
Pork (fresh, chilled, or frozen)	259.0	.25
Bacon and Ham	1.6	—
Sausages	1.3	—
Total Merchandise	143.300.4	

[1]Commodity data: FAO, *Trade Yearbook 1981*; total trade data: IMF, *Balance of Payments Statistics*, Vol. 33, Part I, 1982.

protection is provided by Table 14, showing average domestic prices, import levies, and FOB export prices for beef, pork, and barley during the years 1981-1983. Particularly in the case of beef, the import levies assessed on meat are very high in proportion to domestic prices. Levies exceed 70 percent of domestic prices for beef and 50 percent of domestic pork prices in a carcase basis. Similarly, the prices Sweden receives for its meat exports are well below those prevailing in the domestic meat market, averaging less than 70 percent of domestic prices in recent years. Producer prices for barley are closer to international market prices than in the case of meat, although they still exceed the prices available in export markets by a significant margin.

Although not explicitly export-oriented, Swedish policies have nevertheless resulted in large exportable surpluses of grain and meat. These surpluses have generally required export subsidies on the world market, which are financed

Table 13

Sweden. Foreign Exchange Rate,
Kronor per U.S. Dollar[1]

Year	Average annual kronor per dollar
1983	7.672[2]
1982	6.284[2]
1981	5.032
1980	4.227
1979	4.285
1978	4.514
1977	4.469
1976	4.354
1975	4.141
1974	4.433
1973	4.358
1972	4.762
1971	5.115
1970	5.173

[1]Source (1970-81): *FAO Trade Yearbook,* 1981.

[2]Source (1982): *Federal Reserve Bulletin* 70 (3): A66.

Table 14

Sweden. Domestic Prices, Import Levies, and F.O.B. Export Prices for Beef, Pork, and Barley, 1981-83

		Beef (carcases)			Pork (carcases)			Barley		
		1981	1982	1983¹	1981	1982	1983	1981	1982	1983
Domestic Price² (wholesale for meat, producer for barley)	kr/ton	16,040	18,740	16,220	12,000	14,920	16,220	959.2	1,085.1	1,164.2
Import Levy² (as of 7/1)	kr/ton	11,700	13,200	16,500	5,470	8,300	8,850	3.1	4.2	5.5
F.O.B. Export Price³	kr/ton	11,095	12,141	13,991	8,664	9,930	NA	830.6	1,046.7	863.6

[1]1983 data are for first 10 months only.

[2]Source: *Jordbruksekonomiska meddelanded*, various issues.

[3]Source: estimated from Sweden trade statistics.

largely by payments from the producers' cooperatives supplemented by revenues from import levies. Because the producers themselves bear a large share of export costs, surpluses marketed abroad have the effect of reducing farm income, providing some incentives to farmers to curb protection.

High producer prices were passed on directly to Swedish consumers until 1973. At that time, accelerating inflation of basic foodstuff prices spawned a small consumer protest movement. In response, the government froze the retail prices of a number of key foodstuffs, including beef and pork, and began financing increases in producer prices directly from the treasury. After the retail price freeze was lifted, government subsidies continued in effect for a number of food products. Meat subsidies were reduced after 1980 in an effort to cut government expenditures, and were eliminated entirely at the end of 1983.

The elimination of government meat price subsidies holds important implications for the Swedish meat sector. Rapid wholesale price increases, which translate into similarly large increases at the retail level, suggest further declines in meat consumption. This in turn implies further growth in surpluses, and thus in subsidized meat exports. Assessments to farmers to finance the exports will result in reductions in farm income.

Swedish policy makers have taken a number of steps aimed at bringing meat production into closer balance with consumption. Slaughter premiums on young animals have been instituted in an effort to bring livestock to market at lighter weights and to reduce numbers of breeding animals. Special supports for some small producers of meat animals have been reduced or withdrawn altogether. Efforts to stimulate meat consumption have also begun, such as increasing subsidies on meat products included in school lunches.

The problems currently posed by the meat sector are likely to remain a serious dilemma for future Swedish food and agricultural policy. Within the context of its overall goals for national food security and the maintenance of a healthy social and economic structure in agriculture, the government must attempt to equilibrate the interests of producers in maintaining farm income at a level comparable to that of other workers with those of consumers seeking some degree of protection from high retail meat prices. The abandonment of subsidies as a means of reconciling these interests at the expense of taxpayers makes the search for effective alternative policy instruments more pressing still.

Given this set of priorities, the character of future policy is likely to continue largely along present lines. The goal of self-sufficiency in food production will be retained, although increased attention may be focused on the role of imported agricultural inputs such as fuel and fertilizer in influencing the degree of vulnerability of Swedish agriculture. High producer prices for grain and meat are likely to be retained as a component of national security, social, and regional policies, although efforts to bring meat production more in line with consumption and reduce exports will probably be expanded. Continuing grain surpluses are likely to be tolerated, however, due to the availability of opportunities to export grain commercially and as a component of foreign assistance efforts.

Prospective Impact of Isolated Soy Protein Food Ingredients on Sweden's Domestic Meat Market

Moderating the effects of high producer prices for meat at the retail level has been a central consideration in Swedish food and agricultural policy during the past decade. This goal was formerly pursued through the use of consumer price subsidies; budgetary pressures have now added impetus to the search for alternative measures to reduce the cost of meat products to consumers while maintaining producers' incomes at the desired levels.

Projected Meat Product Prices Under Alternative Processing Technologies

Tables 15 and 16 detail the protein ingredient proportions used in producing two sets of comparable meat products - in each case with one using traditional ingredients, and the other utilizing isolated soy protein. The assumption that the traditional and reformulated products are equal in taste, appearance, and nutritional quality is fundamental to this analysis. It follows from this that at a given price, consumers would be expected to be equally receptive to meat products processed using either technology, and thus that price comparisons between traditional and reformulated products will be meaningful. Other assumptions regarding costs are also important. It is assumed for the purposes of this analysis that processing and marketing costs per kilogram of final product are equal for traditional and reformulated products, and that costs per unit of final product for non-protein ingredients (corresponding to nonmeat ingredients in traditional formulations) are equal for both products. Thus any savings in ingredient costs with the use of isolated soy protein are assumed to be fully reflected in retail meat product prices.

Tables 15 and 16 also present projected ingredient costs for the protein blocks used in traditional and reformulated products. These costs were calculated from 1983 average wholesale carcase prices as an approximation of Swedish industrial meat costs. Isolated soy protein is assumed to cost meat processors 20 kronor per kilogram. Water, salt and other seasonings are assumed to have minimal impact on final product costs and have been excluded from the analysis. Prices for pork fat used in reformulated sausage products were unavailable and were excluded.

The augmentation of processed meats (sausages) with soy protein isolate is projected to result in a 31 percent increase in final product yield from a given meat block input (Table 15). Protein block ingredient costs per unit of final product are estimated to decline by approximately 1.6 kronor per kilogram. Adding to this the additional savings at the retail level due to the retailer's markup over wholesale (assumed at 65 percent), the retail cost of the reformulated product is projected at 2.7 kronor per kilogram less than a

Table 15

Protein Ingredient Proportions for Traditional and Reformulated Meat Products and Projected Cost Savings with the Use of Soy Protein Isolate

Processed Meats (Sausages)	cost/unit (kr/kg)	Traditional		Reformulated	
		quantity (kg)	total cost (kr)	quantity (kg)	total cost (kr)
Meat block (22% fat)[1]					
-60% beef	22.26	22.8	507.528	17.4	387.324
-40% pork	20.28	15.2	308.256	11.6	235.248
Soy protein isolate	20	0	0	1.5	30
Water	0	—	—	6	—
Pork fat	N/A	—	—	1.5	—
Total Products		100	815.784	100	652.572
Projected cost of protein block per kg of final product produced (KR)			8.15784		6.52572
Savings with soy protein (kr/kg)				1.63212	
Savings projected to the retail level (kr/kg)[2]				2.692998	
1983 Average retail price[3] (traditional product) (kr/kg)				38.06	
Total reduction in retail price projected with use of soy protein isolate (percent)				7.075665	

[1]Meat costs approximated from 1983 average wholesale carcase prices (Table 4) using the following adjustment factors: beef, 110% of wholesale carcase price; pork, 125% of wholesale carcase price.

[2]Average markup from wholesale to retail price for mixed cured meat products, December 1983. Wholesale price was 23.00 kr/kg, retail 38.06 kr/kg. Source: data provided by Hans Erik Uhlin, Swedish University of Agricultural Sciences, private communication.

[3]Average retail price for mixed cured meats, December 1983. Source: Hans Erik Uhlin, Swedish University of Agricultural Sciences, private communication.

Table 16

Protein Ingredient Proportions for Traditional and Reformulated Meat Products and Projected Cost Savings with the Use of Soy Protein Isolate

Cooked Hams	cost/unit (kr/kg)	Traditional		Reformulated	
		quantity (kg)	total cost (kg)	quantity (kg)	total cost (kr)
Green ham[1] (with bone & shoulder)	23.17	100	2317	100	2317
Soy protein isolate	20	0	0	1	20
Water and salt	—	17	—	16	—
Total Product (after cooking)	—	100	2317	105	2337
Projected cost of protein block per kg of final product produced (kr)			23.17		22.25714
Savings with soy protein (kr/kg)				.9128571	
Savings projected to the retail level (kr/kg)[2]				1.551857	
1983 Average retail price[3] (traditional product) (kr/kg)				48.5	
Total reduction in retail price projected with use of soy protein isolate (percent)				3.199705	

[1] Meat cost approximated from 1983 average wholesale carcase price (Table 4) (carcase price divided .7).

[2] Average markup from wholesale to retail price for ham with bone, December 1983. Wholesale price was 28.43 kr/kg, retail 48.50. Source: data provided by Hans Erik Uhlin, Swedish University of Agricultural Sciences, private communication.

[3] Average retail price for ham with bone, December 1983. Source: Hans Erik Uhlin, Swedish University of Agricultural Sciences, private communication.

comparable traditional product. At the 1983 average retail price for blended sausages of 38.06 kronor per kilogram, this suggests a retail price reduction on the order of 7.1 percent.

Table 16 presents a similar analysis for cooked hams. Addition of soy protein isolate is projected to result in a 5 percent extension of the green ham input, and to reduce product ingredient costs by about 91 ore per kilogram. After the effect of the retail margin is added, retail prices for reformulated hams are projected as 3.2 percent lower than prices for comparable traditional products.

Prospective Impact on Consumption of Fresh Cuts Versus Processed Meat Products; Implications for Livestock Products

Reductions in retail prices for processed meat products due to the use of isolated soy protein food ingredients are likely to have some effect on consumers' purchasing habits as they choose between fresh and processed meat products. All other things being equal, introduction of soy protein ingredients might be expected to result in some decrease in consumption of fresh meat cuts as consumers shift toward lower priced processed meats and hams. To the extent that this occurs, Sweden's exports of fresh cuts and carcases may increase somewhat as a result. This in turn might require increased expenditures on the part of farmers to finance exports of fresh cuts and carcases.

However, the specific effects of soy protein ingredients on the consumption of various meat cuts, and thus its effects on Swedish livestock producers, are difficult to assess in any detail. For example, the addition of soy protein isolate to meat products would appear to offer possibilities for the Swedish livestock industry to adjust relative meat part prices to allow processed meat products to bear a larger share of total carcase costs. Opportunities would exist to use soy protein as a means for stimulating domestic meat consumption through lower retail prices and by diverting some meat cuts from the export fresh market to the domestic processed market. Some opportunities may also exist for diverting meat exports from fresh cuts into more competitive processed products utilizing soy protein, thus lifting off the domestic surplus with a lower total expenditure on export subsidies. Soy protein may thus prove to be an important tool in efforts to manage the utilization of Sweden's meat production.

Prospective Impact of Isolated Soy Protein Food Ingredients on Sweden's Foreign Trade and Balance of Payments

Isolated soy protein food ingredients, by increasing the output of processed meat products available from a given livestock input and reducing ingredient costs for the reformulated product, present Sweden with an opportunity to increase export earnings from processed meat products while reducing the differential between domestic and international prices for processed meat products.

Table 17 presents estimates of the net effects on Sweden's trade balance of the incorporation of soy protein into processed meat products exported. As in the analysis of the domestic Swedish market, it is assumed that products containing soy protein are equivalent in all respects to traditional meat products, and thus that both traditional and reformulated products would command similar prices in foreign markets. Mixed cured meats (including sausages) and unmixed cured meats (including hams) are the two groups of products examined in this analysis. Ingredient proportions for reformulated products are assumed to be the same as in Tables 15 and 16.

In evaluating the trade impact of isolated soy protein, it is assumed that Swedish meat production remains constant at current levels, and that exports of meat carcases and fresh cuts remain fixed at the 1982 level. Thus, any trade effects are confined to increases in processed meat product exports and soy protein imports.

Total domestic consumption of the two groups of products was estimated for 1982, based on the traditional product formulations employed in Tables 15 and 16. Projected retail price reductions for each group of reformulated products, estimated in Section II, were then used to estimate increases in domestic demand for meat products resulting from the introduction of soy protein ingredients. Augmentation of domestic supplies was then estimated for each product, again utilizing the same formulations employed earlier. The resulting net increase in the supply of exportable meat products was then calculated for each group of products. Exportable surpluses of mixed cured meats are projected to increase by 68,000 tons, and ham exports are projected to rise by about 880 tons.

The value of meat product exports was estimated using three-year average export prices (f.o.b.) calculated from Swedish trade statistics. Mixed cured meat exports are valued at 9810 kronor per ton and unmixed cured meats are valued as 12623 kronor per ton. Import costs for soy protein isolate are evaluated using a c.i.f. price of 19000 kronor per ton. Using these prices and the quantities estimated earlier, the total trade gain to Sweden resulting from the use of soy protein ingredients is projected as approximately 576 million kronor annually.

As the figures in Table 17 indicate, the addition of isolated soy protein ingredients to Sweden's processed meat products appears to offer significant trade advantages. Processed meat products could become a significant contributor to Sweden's export earnings while at the same time increasing the supply of meat products available for domestic consumption. Reduced ingredient costs for reformulated products would also bring Swedish prices for these products more in line with those on the world market, thus reducing or even conceivably eliminating the need for export subsidies on these products. This in turn would suggest opportunities for Swedish livestock producers to reduce their overall export subsidy burden through increased emphasis on more competitive reformulated processed meat products.

Table 17

Projected Impact of Addition of Isolated Soy Protein Food Ingredients on Swedish Domestic Consumption of Processed Meat Products and on Foreign Trade Balance

	Mixed cured meats (includes sausages)	Unmixed cured meats (includes hams)
1982 Domestic Consumption[1] (carcase wt., thousand tons)	140.6	27.7
Proportional meat content of traditional product (assumed)	.38	1
Estimated 1982 consumption of total product (thousand tons)	259[2]	27.7
Percent reduction in average retail prices with addition of soy protein (estimated in Tables 15 and 16)	-7.1	-3.2
Price elasticity of demand[3]	-.67	-.57
Projected increase in domestic consumption with use of soy protein ingredients (thousand tons)	12.320	.505
Projected augmentation of domestic supply with use of soy protein ingredients (thousand tons)	80.379	1.385
Net change in supply available for export (thousand tons)	68.058	.879
Export price[4] (three-year average, F.O.B., kr/ton)	9810	12623
Value of added meat products exports (million kr)	667.655	11.105
Soy protein isolate required (thousand tons)	5.090	.277
Import price (C.I.F., kr/ton)	19000	19000
Cost of soy protein imports (million kr)	96.723	5.263
Net impact of soy protein on trade balance (million kr)	570.932	5.842
Total trade impact, both products (million kr)	576.774	

[1] See Table 8.

[2] Assuming 70 percent meat block yield from carcase equivalent consumption and meat content of 38 percent in final product.

[3] Demand elasticity for sausages derived from beef and pork elasticities presented in L. Furulund, "Subventionernas effekter för jordbrucket och konsumenterna," *JEM*, 1983 (2): 60. Elasticity used for ham is that for all pork products, presented by Furulund.

Prospective Impact of Isolated Soy Protein Food Ingredients on Resource Use in Swedish Agriculture

Swedish food and agricultural policy reflects a number of other considerations in addition to price, budgetary, and income goals. Other goals revolve around national security concerns for limiting reliance on foreign sources of agricultural inputs and products in order to safeguard the food supply from distribution in an emergency. This section examines the prospective impact of soy protein on the use of some important agricultural resources, and discusses the implications for food and agricultural policy goals.

Prospective Impact on Resource Use

The major use of agricultural resources in meat production is for feedgrains for animal feed. By reducing the proportion of animal products contained in meat products, soy protein isolate offers savings in the inputs required for the ultimate production of meat products. This analysis concerns the current use of land, fertilizer, and fuel resources in Swedish feedgrain production, and projects savings in the resources needed to maintain the current consumption of meat products by using soy protein ingredients. This section focuses on pork as the livestock ingredient in processed meat products, as hogs are the major animal produced exclusively for meat in Sweden. Resource savings are projected for varying proportions of hog carcases used for production of processed meat products (sausages).

Savings in resources used in the production of hog carcases were projected using Swedish data on average conversion factors in pork and grain production. Hog production is assumed to require 4 tons of barley per ton of carcase weight. Grain (barley) production is assumed to require .287 hectares of land, 25.8 liters of fuel (including fuel for the harvestor), and 36.5 kilograms of fertilizer per ton of grain produced.

As in previous analyses, traditional processed meat products are assumed to contain 38 percent meat and the remainder other ingredients. Reformulated products containing soy protein isolate are assumed to contain 29 percent meat. The proportion of hog carcases used for production of processed meats is allowed to vary from 30 percent (corresponding roughly to current practice) to as high as 40 percent. Savings due to the use of soy protein ingredients are estimated as the difference in resource use between traditional and reformulated products at each level of carcase utilization.

The results of this analysis are presented in Table 18. At the current level of carcase utilization in processed products, introduction of soy protein ingredients is projected to result in a savings of nearly 41,000 tons of pork (carcase weight), while maintaining total consumption of meat products (both fresh and processed, with processed products including nonmeat ingredients) at the 1982 level. This translates into a savings of nearly 164,000 tons of grain which could be used for other domestic purposes or for export, or, if grain production is reduced by that amount, in a substantial savings in the

Table 18

Projected Savings in Resources Needed to Maintain Domestic Pork Products Consumption Under Alternative Carcase Cutting Practices and Processed Meat Formulations

Meat Product Formulation	Traditional (38% meat)			Reformulated (29% meat)		
Proportion of hog carcase used for processed products	.3	.35	.4	.3	.35	.4
Total projected consumption of pork products (includes nonmeat ingredients, tons)[1]	302275	302275	302275	302275	302275	302275
Percent of total pork product consumption in processed form (% of total product wt., fresh cured & processed meats)	.530	.586	.636	.596	.649	.696
Pork production in carcase wt. needed to maintain total projected consumption of pork products (tons)	289915.4	274861.2	261293.2	248962.7	232550.1	218167.6
Grain required for pork production (tons)[2]	1159662	109945	1045173	995850	930200	872670
Resources required to support grain production for hog feed:						
land (hectares)[3]	333236.1	315932.4	300337.0	286164.0	267298.9	2507067.4
fuel (liters)[4]	29919274	28365677	26965460	25692946	23999168	22514898
fertilizer (tons)[5]	42327.66	40129.74	38148.81	33952.31	31852.47	

Savings With Use of Reformulated Product:			
Pork (carcase wt., tons)	40952	42311	43125
Grain (tons)	163811	169244	172502
Land (hectares)	47072	48633	49569
Fuel (liters)	42263829	4366509	4450562
Fertilizer (tons)	5979	6177	6296

[1]Figure for total domestic consumption of pork products (fresh and cured meats and processed meat products containing pork) calculated as follows:

a. 1982 domestic pork consumption (carcase weight) = 272,733 tons (Table 2).

b. It is assumed that the meat yield of a hog carcase (excluding bones and other inedibles) = 70% of carcase weight, and that under current carcase cutting practices. 30% of the meat yield from hog carcases is used for processing.

c. Assuming that all processed pork products are consumed domestically (which corresponds very closely to the actual situation), the total quantity of pork meat available for processing based on 1982 production of 325,020 tons (Table 1) = .7(.3) (325,020) = 68,254 tons.

d. Assuming an average meat content of 38 percent by weight in traditional processed pork products, total domestic consumption of processed pork products (including nonmeat ingredients) = (68,254/.38) = 179,616 tons.

e. Assuming that the remainder of pork carcase consumption took the form of fresh and cured meats, total domestic consumption of fresh and cured pork = .7 (272,733) - 68,254 = 122,659 tons.

f. Total consumption of pork and pork products (including nonmeat ingredients in processed products) = 179,616 + 12,659 = 302,275 tons.

[2]Assuming feed conversion ratio of 4 kg. grain per kg. carcase weight produced in farrow to finish production. Approximate average of several Swedish sources.

[3]Assuming average grain yield of 3.48 tons per hectare. Source: Swedish Agricultural Market Board.

[4]Assuming average fuel inputs of 25.8 liters per ton of grain produced. Source: Swedish Agricultural Market Board.

[5]Assuming average fertilizer inputs of .0365 tons per ton of grain produced. Source: Swedish Agricultural Market Board.

resources needed for grain production. Savings are estimated at approximately 47,000 hectares of land, 4.23 million liters of fuel, and 6,000 tons of fertilizer.

As the proportion of hog carcases devoted to the production of processed meat products is increased, the potential resources savings become even more significant. If 35 percent by weight of each hog carcase is used in production of processed meat products, use of soy protein ingredients is estimated to offer ultimate resource savings of 169,000 tons of grain or 49,000 hectares of land, 4.31 million liters of fuel, and 6,200 tons of fertilizer. If the carcase cutting ratio is increased to where 40 percent of each carcase is used for processed products, soy protein is projected to result in a savings of 173,000 tons of grain or 50,000 hectares of land, 4.45 million liters of fuel, and 6,300 tons of fertilizer.

The above estimates of resource savings are based on the current average intensity of capital inputs (fuel and fertilizer) in Swedish feedgrain production. Reduced need for feedgrains for meat production (while maintaining grain exports at current levels) would also offer opportunities to use land resources less intensively, thus in effect substituting some or all of the savings in agricultural land resources for even greater savings in fuel and fertilizer supplies. This may be particularly important for Sweden, due to policy goals for keeping arable farmland in production while at the same time lessening reliance on foreign sources of critical agricultural inputs.

Implications for Policy Goals

The results of this analysis suggest potentially large savings in agricultural resources with the introduction of soy protein food ingredients. In the event of an emergency, total consumption of meat products in Sweden could be maintained at a considerable reduction in the livestock inputs required under traditional product formulations. This in turn offers savings either in grain resources needed to support meat production, freeing grain supplies for other uses, or in the resources needed to support grain production. Scarce land, fuel, and fertilizer inputs could then be used to produce food products for direct human consumption.

Should Sweden be subjected to disruption of its foreign supplies of fuel and fertilizer, on which it is heavily dependent, stocks of soy protein isolate could serve as an important asset for maintaining adequate levels of protein consumption. As such it could contribute to Sweden's strategy for ensuring national food security. While soy protein isolate would also have to be imported from abroad, the relative quantity required to substitute for the supplies of fuel and fertilizer needed to support a given level of meat products consumption is quite low. For example, with 30 percent of each hog carcase used for processing, reformulated products would require approximately 2,700 tons of soy protein isolate, balanced against over four million liters of fuel and nearly 6,000 tons of fertilizer necessary to support the same level of meat products consumption using traditional ingredients. The stability of price and supply would also be important considerations in evaluating the relative merits of alternative food system inputs. In this regard, soy protein isolate would appear to offer substantial advantages in comparison to petroleum and petroleum-derived fertilizers.

Summary and Conclusions

This paper has examined the prospective economic impacts of the introduction of isolated soy protein food ingredients in Sweden. Attention was focused on the likely effects of the new technology on Sweden's trade and balance of payments and on some prominent goals of domestic food and agricultural policy. Soy protein isolate appear to offer significant opportunities to increase exports of meat products and improve the country's trade performance. Projected reductions in the cost of reformulated meat products suggest benefits to Swedish consumers. And although the maintenance of the current level of producer prices for livestock products was assumed throughout, the analysis also suggests that Swedish livestock producers may realize benefits as well. Reduced prices for processed meat products would lead to increases in domestic consumption of these products, and improved competitiveness of Swedish meat products on the world market could reduce the need for export subsidies financed by farmers.

Selected Bibliography

Marshall H. Cohen (1975): *New Directions in Swedish Agricultural Policy.* USDA, ERS, Foreign Agricultural Economic Report, No. 104.

——————————————— (1982): *Sweden's Agricultural Policy.* USDA, ERS, Foreign Agricultural Report.

OECD (1980); *Review of Agricultural Policies in OECD Member Countries, 1979* (Country Note on Sweden). Paris: OECD.

Carlos C. Sanchez (1983): *Swedish Farming and the Agricultural Cooperative Movement in Sweden: Structure and Developments.* Stockholm: Federation of Swedish Farmers.

Sweden. Statens Jordbruksnamnd. *Jordbruksekonomiska Meddelanden,* various issues (National Agricultural Market Board, *Journal of Agricultural Economics*).

Chapter 4

Grain and Meat in China: Trends in Consumption, Production, and Imports with Special Reference to Isolated Soy Protein Meat Ingredients

*Terry Sicular**
with
Vanessa Weiss

Food Research Institute
Stanford University

We wish to acknowledge the help of Professor Walter P. Falcon, who provided considerable guidance and helpful comments during the preparation of this manuscript. Financial assistance from the Ralston Purina Company and the Center for Economic Policy Research at Stanford University is gratefully acknowledged. Thanks also Robin A. Cowan for his help in our struggles with the micro computer.

*Terry Sicular is Assistant Professor, Food Research Institute; Vanessa Weiss received her M.A. from the Food Research Institute.

Abstract

China recently embarked upon a new economic agenda to provide higher incomes and more and better consumer products to improve the standard of living of the Chinese people.

Under the new programs, between 1978 and 1981 national income rose 9 percent and consumption 13.5 percent a year. As a percentage of national income, consumption rose from 64 percent to 72 percent.

As personal incomes rise, demand for agricultural products for food also rises. At higher income levels, people also tend to change the composition of their diets; demand for animal products such as meat, poultry, eggs and milk rises faster than demand for grain-based foods.

These trends have been observed in China. If personal incomes grow as planned, per capita consumption of grain and meat in China should continue to rise through the 1980's.

Accommodating these increases and changes in food demand carries major implications for China's grain and livestock sectors. Current trends will generate a growing imbalance between domestic supply of and demand for these products. To achieve balance, the Chinese government will have to implement supply-side policies either to increase imports or to accelerate grain production and improve the efficiency of animal husbandry.

Imports to solve the problem probably would be in the form of grain to free more domestic grain supplies for animal feed. It may make economic sense, however, for China to import isolated soy protein meat ingredients instead. Importing isolated soy protein could save foreign exchange and could help solve several other problems in food production and distribution and in institutional feeding.

Changes in government policy already have stimulated remarkable growth in meat production in China. Between 1978 and 1982, total meat production rose from 8.6 to 13.5 million tons—an average annual growth rate of 12 percent. Although the supply of meat in cities has increased sufficiently to eliminate meat rationing, meat supply still falls short of demand.

Pork currently accounts for 94 percent of China's red meat consumption. Plans call for increased production of ruminant meat—beef, goat and mutton. Chicken now represents a very small portion of the supply. Chinese planners target red meat production to reach 18 to 19 million tons in 1990.

Importing isolated soy protein as an ingredient in meat products could substitute for some of the increase in domestic meat production. This would mitigate rising feedgrain demand and save significant amounts of foreign exchange and probably some domestic resource costs as well.

Isolated soy protein now is economically attractive for China, however, only in comparison to the alternative of importing grain to produce red meat. It may not be so attractive in comparison to producing more poultry.

In addition to the foreign exchange implications, however, isolated soy protein may appeal to Chinese planners because it could save scarce domestic resources used in transportation, grain-handling facilities, and refrigeration and other storage.

Although China produces canned meats and sausages, its meat-processing industry is underdeveloped and is probably not now in a position to make effective use of isolated soy protein meat ingredients on a large scale. The Chinese government recognizes the growth potential of this industry, however. As China's food processing industries develop, the opportunities for isolated soy protein use in processed meat products will increase.

Introduction

The change in political leadership following the death of Mao Zedong and the downfall of the "Gang of Four" in China in the late seventies led to a change in the government's economic agenda. Included in the new agenda was a commitment to provide higher incomes and more and better consumer products in order to improve the living standard of the Chinese people. The Chinese government's commitment to improving living standards has been reflected in a number of recent policy initiatives. One such initiative has been in the area of pricing policy. Substantial increases in the prices the state pays farmers for their products have had a direct and positive effect on rural incomes and thus on rural living standards. Although the state concurrently raised urban retail prices of nonstaple foods, retail price increases have not kept pace with agricultural procurement prices. Thus the government has raised rural incomes while protecting urban consumers. A second set of initiatives has raised urban wages. Widespread grade promotions and the expanded use of bonus awards have enhanced the incomes of workers and staff in state enterprises.

Due to these and other policies, national income and consumption in nominal terms grew 9% and 13.5% per year, respectively, between 1978 and 1981. As a percentage of national income, consumption rose from 64% to 72%. (See Table 1.) The increased share of consumption may, however, be overstated due to relatively rapid increases in consumer prices. With respect to the future, the state has planned for continued growth in consumption and personal incomes, although at slightly lower rates. During the Sixth Five-Year Plan (1981-1985), the state plans to increase total urban and rural consumption by 22%, or approximately 4% a year. Furthermore, the per capita incomes of peasants (who account for 80% of China's population) are targeted to grow 6% a year, and the total wage bill of state and urban collective enterprises is targeted to grow 5% a year. These targets are apparently in real terms, since the Sixth Five-Year Plan calls for stable prices.

The recent and planned future emphasis on consumption and personal income carries implications for China's grain and livestock sectors. As personal incomes rise, so does the demand for agricultural products used for food—grains, meat, poultry, and milk. Moreover, at higher income levels, the composition of the diet changes: demand for animal products (meat, poultry, eggs, and milk) increases more rapidly than the demand for grain-based foods.

Can China's production of grain and animal products keep pace with the rising demand implied by the current policy program? Are the Chinese government's planned targets for future grain and meat production consistent with probable increases in their demand? If not, China will have to address a growing imbalance between domestic supply of and demand for these products. In order to bring about consistency, the Chinese government will have to implement supply-side policies that either permit increased imports or accelerate grain production and improve the efficiency of animal husbandry.

Alternatively, the government can dampen demand by slowing growth in personal incomes, allowing food prices to increase, or rationing supplies of grain and meat.

If the Chinese government uses imports to close the gap between supply and demand, the majority of those imports will probably be in the form of grain. As demonstrated below, however, it may make economic sense for the government to import isolated soy protein meat ingredients instead of grain. At the

Table 1
National Income and Consumption
(billion yuan, current prices)

	National Income	Consumption	As a % of National Income
1957	93.5	70.2	75%
1965	134.7	98.2	73%
1975	245.1	162.1	66%
1978	297.5	188.8	64%
1979	335.6	219.5	65%
1980	368.4	251.9	68%
1981	384.9	275.9	72%
1982	(424.7)	—	—

Average Annual Growth Rates In
National Income and Consumption

	National Income	Consumption
1957-65	4.7%	4.3%
1965-75	6.2%	5.1%
1975-78	6.7%	5.2%
1978-81	9.0%	13.5%

"-" indicates data not available. Parentheses indicate figure is preliminary.

Sources: *Zhongguo Jingji Nianjian, 1981,* p. VI-9.
Zhongguo Tongji Nianjian, 1981, p. 21.
Beijing Review 26 (19), May 9, 1983, p. II.

Note: Relatively rapid consumer price increases relative to the general price level implies that these data overstate the real share of consumption in national income.

margin, importing isolated soy protein could save the Chinese government foreign exchange and could also help solve several other problems of both food production and distribution as well as of institutional feeding on a mass scale.

The remainder of this paper analyzes the major factors that influence the balance between China's demand for and domestic supply of grain. These factors include: personal income growth per capita; the income elasticities of demand for grain and meat; the efficiency of grain use in production; and the performance of domestic grain production. The extent of excess grain demand or supply depends on the assumptions made with respect to future trends in these factors. The paragraphs below discuss probable trends, and the body of the text and attached tables present several projections of excess grain demand based on alternate sets of assumptions.

Rising Incomes and Consumption

Recent increases in national income and consumption have largely been a reflection of rising personal incomes. Bo Yi-bo, a prominent Chinese economist and policy maker, estimates that 86% of national income newly added between 1979 and 1982 went into "the pockets of peasants, communes, brigades, staff and workers".[1] Accordingly, since 1978 growth in rural incomes and urban wages, shown in Table 2, has accelerated. After two decades of 3% average annual growth, per capita rural income grew annually at more than 10% in both nominal and real terms. Urban wage growth rose from 0% to 8% a year in nominal terms, and to about 4% a year in real terms. Since urban labor force participation also increased over this period, per capita urban incomes rose faster than wages. Per capita urban incomes increased approximately 9% a year in real terms.[2]

In low-income countries, increases in per capita personal income lead to higher consumption of foodstuffs. China has been no exception: recent income growth has been accompanied by higher expenditures on food. Rural household survey data show that between 1978 and 1981 per capita rural income grew 64%, and per capita expenditures on food rose 45%. Urban household survey data show that per capita income rose 47% and expenditures on food 41%.[3] In low-income countries, increases in income are typically accompanied by shifts in diet toward nonstaple foods—meat, poultry, eggs, fish, and vegetables. Once again, China bas been no exception. Between 1978 and 1981, per capita grain consumption rose from 196.5 to 219.2 kilograms or 12%, as compared to a 45% increase in per capita pork consumption. Per capita annual consumption of pork, which accounts for 94% of all red meat consumed in China, increased from 7.7 to 11.1 kilograms (See Table 3.) Consumption of other animal products such as poultry and dairy products has also risen, although per capita consumption levels are still very low. As of 1981 poultry meat consumption per capita remained less than one kilogram and fresh egg consumption less than 2.5 kilograms per year. Milk consumption is also low, in 1980 averaging only 1.4 kilograms per person per year.[4]

Table 2
Urban and Rural Incomes
(current yuan per annum)

	Per Capita Income Distributed by Rural Collectives [a]	Average Annual Growth Rates	Rural Household Survey Per Capita Income [b]	Average Annual Growth Rates	Average Wage Per Worker in State Enterprise [c]	Average Annual Growth Rates	Average Wage Per Worker in Urban Collective Enterprise [c]	Average Wage Per Worker in Both Urban Collective and State Enterprise [c]	Average Annual Growth Rates
1957	40.5		—		637		—	—	
1965	52.3		—		652		—	—	
1970	59.5	3% ('57-'78)	—		609	0% ('57-'78)	—	—	
1975	63.22		—		613		—	—	
1977	64.98		—		602		—	—	
1978	74.00		133.57		644		505	614	
1979	83.40		160.17		705		542	668	
1980	85.90	11% ('78-'81)	191.33	19% ('78-'81)	903	8% ('78-'81)	624	762	8% ('78-'81)
1981	101.32		223.44		812		642	772	

[a] *Zhongguo Nongye Nianjian,* 1980, p. 41; *Zhongguo Jingji Nianjian,* 1981, p. VI-30; *Zhongguo Tongji Nianjian,* 1981, p. 198. Includes income in cash and in kind distributed by rural collectives to their members.

[b] *Zhongguo Tongji Nianjian,* 1981, p. 431. Includes household income distributed by the collective, as well as private sideline and other noncollective income received by households.

[c] Chinese State Statistical Bureau figures in World Bank, *China: Recent Economic Trends and Policy Developments,* p. 121.

Note that average wage per capita will be lower than average wage per worker, depending on the number of dependents per worker.

"—" indicates data not available.

Table 3

Per Capita Grain and Pork Consumption

	Grain Consumption Per Capita [a,e] (kg trade grain)			Pork Consumption Per Capita [b,e] (kg)	Red Meat Consumption Per Capita [c] (kg)
	Urban	Rural	National Average		
1952			197.5	5.91	6.26
1954			195.5		
1956			204.5		
1957	196.0	204.5	203.0		
1958			198.0		
1959			186.5		
1960			163.5		
1962			164.5		
1965			184.0		
1970			188.0		
1972			173.5		
1975			191.5		
1976			182.8		
1977			193.0	7.0	7.4
1978	216.5	192.5	196.5	7.65	8.1
1979			205.0	9.7	10.3
1980			213.5	11.2	11.8
1981			219.2	11.1	11.8

Grain projections: $(\eta_g=0)$, $(\eta_g=.22)$, $(\eta_g=.37)$

Year	$(\eta_g=0)$	$(\eta_g=.22)$	$(\eta_g=.37)$
1985	219.2[d]	231[d]	239[d]
1990	219.2[d]	247[d]	267[d]

Red Meat projections: $(\eta_m=0)$, $(\eta_m=.81)$, $(\eta_m=1.34)$

Year	$(\eta_m=0)$	$(\eta_m=.81)$	$(\eta_m=1.34)$
1985	11.8[d]	14[d]	16[d]
1990	11.8[d]	18[d]	24[d]

[a] All grain consumption figures from Landry, 1983, except projections.

[b] World Bank, *China: Recent Economic Trends and Policy Development*, p. 12.

[c] All numbers calculated assuming pork/meat ratio of 94% except 1990 where the ratio is assumed 92%. USDA, *China: World Agricultural Regional Supplement*, p. 19.

[d] Derived using an average annual growth rate of 6% for income, and alternative income elasticities for grain and red meat. Target growth rates of income given in the Sixth Five-Year Plan were 6% for peasant per capita income and 4.9% for the total wage bill of state and urban collective enterprises. See *Beijing Review*, 25 (51), p. 18.

[e] Data for the late seventies and early eighties, and probably for earlier years as well, include not just rations but also consumption of processed food grain and meat, and consumption supplied by private household production and free market purchases. See *Zhongguo Tongji Nianjian*, 1981, pp. 421,439.

The recent rapid growth in personal incomes and consumption follows two decades of stagnant living standards. Between 1957 and 1977, per capita grain consumption actually declined from 203 kilograms to 193 kilograms. Pork consumption rose modestly, from 5.9 kilograms in 1952 to 7 kilograms in 1977. (See Table 3.) Even with recent growth, China's per capita meat consumption levels remain low by international standards. Per capita grain consumption, on the other hand, is relatively high.

By dividing the average annual percentage increases in per capita consumption by the average annual percentage increase in per capita income, crude marginal income elasticities of demand for grain and meat can be calculated. These income elasticities give a rough idea of the relationship between incomes and consumption. (Data are insufficient to calculate more precise elasticities. See Appendix 1.) Between 1978 and 1981, the crude income elasticity of demand for grain calculated using nominal incomes was 0.22, and for meat 0.81. The elasticities thus calculated are comparable to those in other Asian less-developed countries.[5] Calculated using real income, the crude income elasticity of demand for grain was 0.37 and for meat 1.34, relatively high numbers.

These crude elasticity estimates include not just income effects, but also the effects of changing prices and supply availabilities. Over this period, the prices paid for meat and grain by the majority of consumers rose. In urban areas, the state retail sales price of red meat rose about 30% in nominal terms, or about 16% in real terms. Nominal state grain retail sales prices remained unchanged, and so declined about 11% in real terms. In rural areas, where 80% of China's population resides, consumer prices of both grain and meat rose faster than the cost of living. In nominal terms, meat prices rose about 30% and grain prices 45%. In real terms, their price increases were roughly 5% and 18%, respectively.[6] All else held equal, these price increases should have dampened the demand for grain and meat, thus making the "true" income elasticity estimates larger than those cited above.

Concurrently, the supplies of grain and meat available to consumers increased. State rationing and suppression of private exchange had in the past held consumption below desired levels. During recent years such quantity restrictions have been eased. In urban areas, meat rationing has been relaxed and above-ration grain has become available on free markets and through state negotiated sales. In rural areas, state-planned grain supplies have increased. Rural free markets have also expanded, improving the availability of these and other commodities. Even without the rise in incomes, improved availabilities would probably have led to higher consumption levels. Thus improved availabilities have reinforced the positive income effects.

The income elasticities presented above are not adjusted to control for price and supply availability effects. Ignoring increases in grain and meat prices would cause the elasticity estimates to understate their true values, while ignoring rising supply availabilities would cause them to overstate the true values. Depending on the relative strength of these two effects, the elasticity estimates could understate or overstate the true relationship between per capita income and consumption.

If personal incomes grow as planned, per capita consumption of grain and meat in China should continue to rise through the 1980s. The extent to which consumption will rise in response to income increases depends on the income elasticities of demand. If the income elasticity of grain remains constant at 0.37 and that of meat at 1.34, and if per capita income grows at the target rate of 6%, per capita direct grain demand would increase from 219 kilograms in 1981 to 239 kilograms in 1985 and 267 kilograms in 1990. Per capita meat demand would rise from 11.8 to 16 to 24 kilograms. Although these meat demand levels are believable, the projections for per capita grain consumption are extremely high. It is more likely that the income elasticity for grain would decline somewhat as income increases and that per capita grain consumption would plateau, as it has in other countries. Predicted per capita consumption levels calculated using the lower, nominal income elasticities of 0.22 for grain and 0.81 for meat are more reasonable. By 1990 per capita grain consumption would then reach 247 kilograms and meat consumption 18 kilograms. (See Table 3.) Since these per capita consumption levels appear to be more realistic than those calculated using the real income elasticities, in the remainder of this paper it is assumed that the lower, nominal income elasticities apply (unless noted otherwise).

Levels of per capita consumption directly influence total grain and meat demand. When multiplied by total population growing at planned rates (see Table 4), the higher set of real income elasticities implies a 1990 demand for grain as human food of more than 360 million tons and a demand for meat of more than 27 million tons. Production of 27 million tons of meat would require about 115 million tons of feedgrain.[7] Including additional grain fed to poultry and dairy cows, grain used for seed and industrial purposes, and grain lost through waste, by 1990 total grain disappearance would reach 526 million tons. If the income elasticities were lower, 0.22 for grain and 0.81 for meat, total grain disappearance would reach 471 million tons by 1990. Even in the unlikely case that per capita consumption levels remain unchanged at their 1981 levels (income elasticities of demand for meat and grain equal zero), population growth alone would raise the total demand for grain to 403 million tons. (See Table 5.) Thus, under alternative income elasticity assumptions total grain disappearance in 1990 exceeds 1981 levels by at least 16% and perhaps by as much as 50%. These projections assume population and income growth as planned and that the grain:meat conversion ratio remains at 4.25. If population or income growth differs from planned levels, or if the grain:meat conversion ratio increases, then total grain and meat demands will follow a different course. (See Appendix 2 for grain disappearance projections under alternate income and conversion ratio assumptions.)

Table 4

Population

	Population[a] (1,000 persons)	Growth Rate per annum
1952	574,820	
1957	646,530	
1965	725,380	
1970	825,920	
1971	847,790	
1972	867,270	
1973	887,610	1.9% ('57-'77)
1974	904,090	
1975	919,900	
1976	932,670	
1977	945,240	
1978	958,090	
1979	970,920	
1980	982,550	1.3% ('78-'81)
1981	996,220	
1985	1,060,000[b]	1.6% ('81-'85)
1990	1,131,000[b]	1.3% ('85-'90)

[a] State Statistical Bureau data.

[b] A Chinese 1985 population target of 1.06 billion and a target growth rate of 1.3% are given in the "Report on the Sixth Five-Year Plan" given by Zhao Ziyang on November 30, 1982, printed in *Beijing Review* 25 (51), pp. 10-20. The 1985 target implies 1.6% annual growth between 1981 and 1985. As the 1.3 % target growth rate is expected to continue through 1990, I use it to calculate the population increase between 1985 and 1990.

Table 5

Grain Disappearance
(1,000 mt original grain)

	Human Consumption[a]	Red Meat Feed[b]	Dairy Feed[c]	Poultry Feed[d]	Seed, Ind., Waste[e]	Total Disappearance
1952	136,780	15,292			16,392	
1957	158,128				19,505	
1965	160,807				19,453	
1970	180,076				23,952	
1972	181,290				24,048	
1978	226,825	32,984	308	1,916	30,477	292,510
1979	239,806	36,200	397	1,942	33,212	311,557
1980	252,740	49,275	445	1,965	32,056	336,481
1981	263,098	49,959	488	1,992	35,502	348,039
	(ηg=0) / (ηg=0.22) / (ηg=0.37)	(ηm=0) / (ηm=0.81) / (ηm=1.34)				(ηg=0)(ηm=0) / (ηg=0.22)(ηm=0.81) / (ηg=0.37)(ηm=1.34)
1985	280,000 / 295,000 / 305,000	53,000 / 63,000 / 72,000	790	2,586	36,000	372,000 / 397,000 / 416,000
1990	299,000 / 337,000 / 364,000	57,000 / 87,000 / 115,000	1,463	3,506	42,500	403,000 / 471,000 / 526,000

[a] Converted to original grain equivalents using a trade grain: original grain ratio of 0.83:1.

[b] Assumes 4.25 kg. grain are required to produce one kilogram red meat.

[c] Dairy feed requirements are estimated using data on the number of dairy cows and on production of cow and goat milk given in *Zhongguo Tongji Nianjian*, 1981, p. 164; *Zhongguo Nongye Nianjian*, 1980, p. 118; USDA Economic Research Service, *China: World Agriculture Regional Supplement, Review of 1982 and Outlook for 1983*, p. 13; and *Zhongguo Nongye Nianjian*, 1981, p. 163. Using data on the number of dairy cows and cow milk production for 1978-81, and assuming dairy cows consume 600 kg. grain per year (Owen L. Dawson, *Communist China's Agriculture: Its Development and Future Potential*, Praeger Publishers, New York, 1940, p. 179), I estimate an average of 0.32 kg. grain was used to produce each kg. cow milk during 1978-81. Milk production is assumed to increase at 13% per annum from 1981 to 1990, which is slightly lower than the average annual growth rate of milk output (13.8%) during 1978-81. Similarly, I use a figure of 30 kg. feedgrain per dairy goat per year (Dawson, p. 179) to estimate a feedgrain requirement of 0.265 kg. per kg. goat milk. Goat milk production is assumed to grow at 14% per annum from 1981 to 1990. From 1970 to 1980 actual goat milk output increased 14%. The goat milk output level was quite volatile in the late 1970's, and it grew at an average annual rate of over 40% between 1978 and 1981.

[d] Data on poultry meat and egg production and consumption is incomplete. National average consumption of poultry meat in 1978 was less than 1 kg. per capita. The agricultural household survey shows per capita rural poultry consumption increasing from 0.25 kg. in 1978 to 0.71 kg. in 1981. Urban household survey data gives per capita urban poultry consumption of 1.92 in 1981 (see footnote 4). It is not clear if these consumption figures are calculated on a carcass or live weight basis. I assume per capita poultry meat consumption (carcass basis) of 1 kg. per capita for 1978-81, thereafter increasing at a rate of 5% per year per capita. This is converted to feedgrain equivalents using a ratio of 2 kilograms grain per kilogram poultry meat. (Fang Yuan, in "Wo Guo Nongye Xiandaihua de Jiben Renwu Yingxti Tigao Danwei Mianji Chanliang," *Jingji Yanjiu*, No. 3, 1980, p. 8, gives a grain: chicken meat ratio of 2:1. See also U.S. Joint Economic Committee, *China Under the Four Modernizations*, Part 1, 1982, p. 457.) Eggs are considered a joint product, requiring no additional feedgrain.

[e] Calculated as 10% of grain production.

Production

For many years, China followed a general policy of food self-sufficiency. Consumption of food was supplied primarily by domestic agricultural production. The self-sufficiency objective led the government to experiment with various strategies for promoting production, especially of grain, so that supply would keep pace with the demand of the growing population. At times, however, production has set an upper limit on consumption levels: quantity rationing ensured that even though demand might exceed domestic supply, realized consumption did not. Domestic production and government policies influencing domestic production thus have carried important implications for consumption as well.

Most red meat has been and continues to be produced privately by farm households. In 1980, 90.5% of all pigs were raised by households,[8] and pork accounts for 94% of red meat consumption. (See Table 6.) Until recently, the state has emphasized collective agricultural production. As a result, meat production has been a secondary (and at times marginal) occupation. Since collective enterprise has employed most available productive resources, the labor devoted to hog raising has been supplied primarily by children or the elderly, and table scraps, waste, and forage have constituted a significant proportion of animal feed. Moreover, during "leftist" periods such as the Great Leap Forward (1958-61) and the Cultural Revolution (1966-76), restrictions were placed on the number of pigs households were permitted to raise. Despite its secondary status, red meat production grew at 3.7% per year between 1957 and 1978. (See Table 6.) Two factors explain this persistent growth: first, production in the 1950s was low, providing a small initial base, and second, hog raising was one of the few dependable sources of cash income to rural households and of fertilizers to the collectives.

With the recent political liberalization, the Chinese government has implemented a number of policies to promote meat production. For a short time in the late 1970s, hog raising was attempted with large-scale, mechanized confinement systems; however, this was soon abandoned as feed costs became too high. Household production, however, has continued to expand. With the responsibility system reforms household enterprises are now both condoned and encouraged. Moreover, the state has allowed larger household private plots, thus permitting more fodder production, and has improved producers' access to feedgrain supplies. Finally, the state has raised hog procurement prices 26%, with added price incentives for heavier hogs.

Lifting restrictions and providing extra incentives have brought about remarkable growth in meat production. Between 1978 and 1982 total meat production rose from 8.6 to 13.5 million tons. (See Table 6.) The average annual growth rate over this period was 12%. The government's policies have been so successful that they have led to an oversupply of hogs: state procurement stands, constrained by slaughter and storage capacity, have not been able to purchase as many hogs as the farmers wish to sell. It is unlikely, however, that the oversupply of hogs has translated into an oversupply of meat

Table 6

Production and Growth Rate of Red Meat

	I Pork, Beef & Mutton Production^a (1,000 m.t.)	Growth Rate per annum	II Pork^c (1,000 m.t.)	Growth Rate per annum	Pork/Meat Ratio	Other Red Meat^f (1,000 m.t.)	Growth Rate per annum
1952	3,385		3,024		89.3%	361	
1957	3,985		3,488		87.5	497	
1965	5,510	3.7%	4,982	4.0%	90.4	528	1.5%
1970	5,965		5,648		94.7	317	
1975	7,970		7,279		91.3	691	
1978	8,563		7,890		92.1	673	
1979	10,624		10,010		94.2	614	
1980	12,054	12.1%	11,341	12.7%	94.1	713	4.1%
1981	12,609		11,884		94.3	725	
1982	13,508	1.2% or 2.6%	12,718	1.1% or 2.5%	94.2	790	2.6% or 4.1%
1985	14,000^b or 14,600		13,146^d or 13,709		93.9	854 or 891	
1990	18,500^b or 18,000	5.7% or 4.3%	16,928^e or 16,470	5.2% or 3.7%	91.5	1,572 or 1,530	13.0% or 11.4%

^a From *Zhongguo Tongji Nianjian*, 1981, p. 163.

^b The first respective targets are reported by JEC, *China Under the Four Modernizations*, Vol. 1, p. 437, and the second ones are from USDA, *China: World Agriculture Regional Supplement*, 1983, p. 18.

^c 1952-1975 numbers derived using number of slaughtered hogs from *Zhongguo Tongji Nianjian*, 1981, and dressing weights from Alan Piazza, "Trends in Food and Nutrient Availability in China, 1950-81," World Bank Staff Working Papers No. 607, 1983, p. 90. 1978-1982 numbers from USDA, op. cit., p. 16.

^d Derived using USDA estimate of pork:meat ratio for 1985 of 93.9%, USDA, op. cit., p. 19.

^e Derived using USDA estimate of pork:meat ratio for 1990 of 91.5%, USDA, op. cit., p. 19.

^f Consists mainly of beef and mutton (goat and sheep meat); derived by difference of columns I and II.

to urban consumers. Although the supply of meat in cities has increased sufficiently to eliminate meat rationing,[9] and in some places the supply of pork fat exceeds demand (as the supply of vegetable oils has risen, consumers have substituted oils for animal fats), meat supply still falls short of demand.[10]

Chinese planners target red meat production to reach 14-15 million tons in 1985, and 18-19 million tons in 1990. These targets imply more conservative annual growth rates: about 2% through 1985 and about 5% from 1985 to 1990. At these planned growth rates, future meat production should be sufficient to meet the demand levels implied by planned growth in personal income and population growth.

In contrast to meat, until the past few years grain has been produced collectively.[11] Moreover, grain production has been considered the primary occupation of the agricultural sector, at times to the detriment of commercial crops and sideline occupations. The state has often relied on production and acreage targets rather than price incentives to promote collective production of grain. Much of planning's emphasis has been on rice, wheat, and corn. Production of these grains has expanded steadily over the past two decades with the extension of irrigation, intensification of cultivation, and the introduction of new seeds. Between 1957 and 1978, total grain production grew at a rate of 2.1% (see Table 7), a respectable rate for a country where grain occupies more than 80% of cultivated area and where cultivatable land is scarce. Since production growth has barely exceeded the population growth rate of 1.9%, however, per capita grain consumption has increased little.

Since 1978 grain production, like meat production, has accelerated. From 1978 through 1982, its average annual growth rate was 3.8%. Increased rates of growth in the past six years have followed adoption of the responsibility system (which allows households more decision-making control and ties monetary rewards to production performance), significant increases in grain quota and above-quota procurement prices, reformulation of acreage targets to conform more closely to regional and local comparative advantage, and liberalization of regulations in private exchange of grain.

Future grain production targets are 360 million tons for 1985 and 425 million tons for 1990. These targets, which were probably set on the basis of 1980 production levels, call for 2.3% annual growth in grain output from 1980 to 1985, increasing to 3.4% from 1985 to 1990. To date, grain production has risen more rapidly than planned, with 1981 output up 14% over 1980, and 1982 up 9% over 1981. Preliminary estimates of 1983 output stand at 370 million tons, a 6% increase over 1982 production.[12] The 1983 level of production actually exceeds the Sixth Five-Year Plan target for 1985.

The 1990 grain production target of 425 million tons requires 2% annual growth from 1983, about equal to historical rates but only half the 4% average annual growth of 1978-82. Recent rapid growth rates have been declining, however, and probably will not be sustained: they reflect one-time gains following institutional and production planning reforms in the late 1970s and early 1980s, as well as a rapid increase in fertilizer availability as 13 imported plants purchased in the early 1970s came on line. Nevertheless, future planned output levels requiring only 2% average annual growth through 1990 appear to be conservative, and grain production in 1990 could very well exceed the target.

Table 7

Grain Production and Growth Rates

	Grain Production[a] (Unmilled Grain) (1,000 m.t.)	Growth Rate per annum
1952	163,920	
1957	195,050	
1965	194,530	
		2.1%
1970	239,520	
1975	284,529	
1978	304,770	
1979	332,120	
1980	320,560	3.8%
1981	325,020	
1982	353,430	
		0.62%
Chinese Targets 1985	360,000[b]	2.3%
1990	425,000[c]	

[a] *Zhongguo Nongye Nianjian*, 1980, p. 34. Includes rice, wheat, corn, sorghum, millet, and other coarse grains, as well as potatoes (converted to grain equivalents on a 4:1 ratio before 1964, and 5:1 ratio from 1964 on), and soybeans (dry weight, out of the pod).

[b] *Beijing Review*, 26 (21), May 23, 1983, p. IX.

[c] Chinese production targets from *China Under the Four Modernizations*, Part I, p. 437.

The Balance Between Supply and Demand of Grain

If grain production grows at planned rates, the total demand for grain implied by planned income and population growth will probably exceed domestic supply. As discussed above, total grain demand includes direct human grain consumption, feedgrain requirements derived from human meat consumption, feedgrain for dairy cows and poultry, plus grain used for seed, industrial uses, and waste. The largest component is direct human consumption, which made up 76% of total disappearance in 1981 and is estimated to decline to about 72% by 1990. Depending on the income elasticity of grain demand, direct human consumption of grain will grow at a rate of 2.8% to 3.7% a year during the 1980s, outpacing the planned rate of growth in grain production. Feedgrain demand, the second largest component, will grow even faster. The feedgrain required to meet estimated red meat consumption will grow at the sum of the rate of increase in meat demand plus the rate of change in feedgrain required per unit meat.[13] If the grain:meat ratio remains constant, then feedgrain requirements will grow at the same rate as estimated meat consumption, or an average 6.4%.to 9.7% a year through the 1980s, far exceeding planned rates of grain production growth. The rapid growth in direct and derived grain demand will lead to an increasing imbalance between total grain disappearance and planned domestic production. By 1990, grain disappearance could exceed planned production by between 45 and 100 million tons. (See Table 8)

The extent of this imbalance depends on five key factors: (1) the rate of growth of personal incomes and thus of per capita demand for grain and red meat, (2) the population growth rate, (3) the performance of China's grain production, (4) the pricing of meat and grain, and (5) the efficiency of China's meat production. The future of personal incomes depends to a large degree on the priority assigned it by Chinese leaders. Since population is planned to grow on average at a rate of 1.4% through the 1980s, in order to sustain 6% real growth in per capita personal incomes either national income will have to grow 7.4% a year in real terms, or leaders will have to accept an increasing proportion of national income going to personal income. Historically, China's national income (constant prices) has grown at an average annual rate of about 5%, and in recent years (1978 to 1981) at about 7%.[14] It is unlikely that real national income growth can be raised to 7.4% and maintained at that level through the 1980s. Thus the targets for personal income growth imply increasing the proportion of personal income in national income, and so decreasing the proportion of national income used for investment and other purposes. In recent years China's leadership has demonstrated a willingness to allow more national resources to go into "private pockets," but future policies remain unclear. If leaders decide to accept slower personal income growth rather than cut into state investment funds, then per capita consumption levels would be lowered and the projected imbalance between grain production and demand reduced. If, for example, annual per capita income growth was 4% instead of the planned 6%, grain disappearance and the resulting imbalance would be reduced by about 7 million tons in 1985 and 25 million tons in 1990. Although these reductions would not eliminate the imbalance, they would shrink it

Table 8

Estimated Net Grain Trade and Stock Changes

	A	B			C			D
	Domestic Production [a]	Estimated Grain Disappearance Under Alternate Income Elasticity Assumptions [a,b]			Estimated Net Trade and Stock Changes [c] (A-B)			Actual Net Trade [d] (exports-imports)
1978	304,770	292,510			12,260			-6,955
1979	332,120	311,557			20,563			-10,705
1980	320,560	336,481			-15,921			-11,811
1981	325,020	348,039			-23,019			-14,116
		I	II	III	I	II	III	
1985	360,000	372,000	397,000	416,000	-12,000	-37,000	-56,000	—
1990	425,000	403,000	471,000	526,000	22,000	-46,000	-101,000	—

Notes: [a] These figures are taken from Tables 5 and 7.

[b] Three alternate projections are given for 1985 and 1990. All projections assume a grain: meat conversion ratio of 4.25. Projection I assumes the income elasticities of both grain and red meat are zero, that is, per capita consumption levels remain at their 1981 levels. Projection II assumes the income elasticity of grain is 0.22 and of meat 0.81. Projection III assumes the income elasticity of grain is 0.37 and of meat 1.34. See the Appendix for a discussion of income elasticities.

[c] Estimated net trade stock changes is equal to domestic grain production minus estimated grain disappearance.

[d] No data is available for actual net stock changes during 1978-81.

substantially: The 1985 grain imbalance would be reduced 20% and the 1990 imbalance by more than 50%. (See Appendix 2.)

The imbalance will also be reduced if grain production grows faster than planned. Grain production in 1982 and 1983 has, in fact, exceeded expectations. The annual growth rate between 1979 and 1983 has averaged 4%. As a result, an average growth rate of only 2% will be required from 1983 to meet the 1990 target. The probability that grain output will grow faster than 2% a year is high. If grain production grew 4% annually from 1981 through 1990, by 1990 China would enjoy a net grain surplus of 16 million tons (see Appendix 2).

The future of China's population growth is difficult to predict. Recent census and State Statistical Bureau data indicate that population is growing at a rate close to the planned average rate for the 1980s.[15] Future population growth rates depend on the success of China's birth and family planning program. In recent years China has taken active measures to control birthrates, encouraging use of birth control and creating strong economic and political incentives to reduce family size. Successful family planning would to some extent mitigate the grain imbalance.

Pricing of meat and grain can influence future trends in both production and consumption. With respect to production, past experience has shown that even strict planning targets cannot completely overcome an inappropriate price structure. At the present, one example of inappropriate pricing is the policy of paying a higher price per kilogram for hogs exceeding certain specified weights. This price incentive policy encourages farmers to raise hogs with a high fat content and to slaughter them at a later age, both undesirable results for reasons to be discussed in more detail below. The price of hogs relative to feedgrains is also important: if the price paid for hogs is high relative to the price paid for grains, then farmers will have an incentive to reduce direct marketing of grain and use the grain to raise livestock. The recent rapid increases in hog production imply a hog grain price spread favoring use of grain for feed. Future price adjustments raising the relative price of grain would encourage direct marketing of grain and discourage livestock production, thus reducing total feedgrain requirements. Total feedgrain requirements could also be reduced if prices of more grain-efficient animals like poultry were raised relative to prices of hogs and beef cattle.

On the consumption side, higher prices for meat and grain would tend to dampen demand. In urban areas grain prices in real terms have declined while meat prices have risen. In rural areas, real prices for both grain and meat have increased. In both urban and rural areas, however, the impact on consumption of higher income and improved availabilities have swamped any price effects.

The final determinant of the grain imbalance is the efficiency of meat production. As mentioned above, the derived demand for feedgrains is the second largest component of total grain disappearance. The derived demand for feedgrains depends not only on the demand for meat, but also on the amount of grain required to produce one kilogram of meat. The impact of livestock feedgrain requirements on total grain demand is discussed in more detail below.

The Efficiency of China's Meat Production and the Grain:Meat Ratio

Chinese livestock production is often considered to be fairly efficient in its use of grain. Western observers have commented that the quantity of grain fed to animals is relatively low. Despite such reports, the average amount of feedgrain required to produce one kilogram of meat is not significantly lower than in other countries. Estimated ratios of grain to meat range from 2.2 to 6.0 kilograms grain per kilogram meat. The actual ratio probably lies in between. Chinese sources estimated in 1980 that feedgrain used about 16% of total grain production, which implies a grain:meat ratio of 4.25. This is comparable to ratios for pork in a number of other less-developed countries.

The reason Chinese livestock production is reported to be grain efficient despite unremarkable grain:meat ratios is because the amount of grain fed to animals on a daily basis is low. The organization of production, where hogs and other animals are raised on a small scale with labor-intensive methods by farm households, economizes on grain. Livestock are fed table scraps, millings, and forage as substitutes for feedgrain. The apparent paradox between this grain-efficient livestock regimen and China's unimpressive feedgrain-to-meat ratio is due to the fact that China's stock of animals is large, but the percentage slaughtered each year is relatively small, only 60%. In developed countries, the slaughter rate exceeds 120%, and in the U.S. the rate is 150%. High slaughter rates indicate that animals are slaughtered at an early age, e.g., hogs at ages of 6 months or less. The practice of slaughtering animals young arises because meat gain per unit feedgrain is highest early in the animal's life. As animals grow older, they continue to eat, but grow more slowly and produce proportionately more fat. At present, hogs in China are often raised to an age of one or two years. As a result, even though each animal's daily grain intake is low, the amount of feedgrain consumed per unit meat produced is not.[16]

Without modernization of animal husbandry practices, as future meat production grows to meet rising demand China's ability to hold down grain:meat ratios will probably diminish. First, as the number of hogs and other livestock raised per household grows, the amount of table scraps and millings available per animal will decline. Households will be forced to turn increasingly to grain as livestock feed. Second, the proportion of livestock raised collectively or by the state in confinement systems will eventually rise. Large-scale meat production will require more grain per unit of meat than household production. On the margin, then, animals' daily feedgrain intake will probably rise.

Two forces, however, counterbalance this trend. First, over time the Chinese plan to develop pastoral regions and so increase the proportion of ruminant meat—beef, goat, and mutton—in total meat production. The USDA estimates that by 1990 the percentage of ruminant meat will increase from 6% at present to 8 or 9% of total meat output. This policy will better utilize China's agricultural resources and help reduce grain requirements. Second, the Chinese are trying to improve the efficiency of meat production methods. Traditional animal breeds and husbandry practices can be modernized. The state is actively promoting genetic research to develop breeds that produce more meat per unit of feed. For hogs, they are concurrently trying to develop and promote breeds that yield a higher proportion of lean meat to fat. The state is also encouraging the expanded use of scientifically formulated feed concentrates composed of grain, carbohydrates, and proteins carefully matched to animal nutrient requirements. The careful matching of animals' feed to their diet requirements should reduce grain waste. Finally, the state is also pushing to increase slaughter rates, promoting reduction of hog lifetimes from one year to six months and sheep lifetimes from two years to one year.[17]

Thus there exist two countervailing forces influencing feedgrain-to-meat ratios: on the one hand, increased numbers of animals per household and the trend toward large-scale meat production will tend to increase the amount of grain required per unit of meat; on the other hand, modernization of production will tend to reduce feedgrain requirements. The net outcome depends on the success of China's modernization drive. At best, the ratio will probably remain at 4.25, and it could edge higher. If the ratio remains at 4.25, indirect demand for feedgrain will grow between 6.4% and 9.7% a year through 1990, faster than planned grain production. If the feedgrain requirements per unit of meat rise, then the imbalance between grain production and demand will be exacerbated. If on the margin the grain:meat ratio increases to 6.0, then the demand for feedgrain in 1990 will be 23 to 46 million tons higher than the projected levels given in Table 7. (See Table 8a.)

China's Trade Policy

Due to China's diversified resource endowment, huge size, and extensive domestic commerce, foreign trade has remained small relative to national income. Moreover, during the fifties and sixties a combination of domestic policies supporting self-reliance and international events affecting China's relations with Western countries hindered the growth of trade. During the 1970s, however, both exports and imports have grown rapidly. The recent rapid growth of China's foreign trade—the total value of imports and exports in current dollars has increased on average 23% a year through the seventies— reflects both improvement in China's international relations and recognition by China's leaders that trade can be used to promote economic modernization.

With the expansion of trade, China's policy of absolute self-reliance has been superseded by the more flexible policy of expanding trade but maintaining the balance of payments. Despite this policy, China incurred trade deficits five out of ten years during the 1970s. These deficits resulted from lagging exports due to the world recession, domestic economic stagnation during the leadership change in the mid-seventies, and uncontrolled growth in imports following

Table 8a

Estimated Net Grain Trade and Stock Changes, With Higher Marginal Grain:Meat Conversion Ratio of 6:1

	A	B	C	D
	Domestic Production[a]	Estimated Grain Disappearance Under Alternate Income Elasticity Assumptions[a,b]	Estimated Net Trade and Stock Changes[c] (A-B)	Actual Net Trade[d] (exports-imports)
1978	304,770	292,510	12,260	-6,955
1979	332,120	311,557	20,563	-10,705
1980	320,560	336,481	-15,921	-11,811
1981	325,020	348,039	-23,019	-14,116
1985	360,000	I 394,000 II 425,000 III 447,000	I -34,000 II -65,000 III -87,000	—
1990	425,000	I 426,000 II 506,000 III 572,000	I -1,000 II -81,000 III -147,000	—

Notes: [a]Except for grain disappearance projections, these data are taken from Tables 5 and 7.

[b]Three alternate projections are given for 1985 and 1990. They all assume a marginal grain:meat conversion ratio of 6:1 (1978-81 figures use a lower ratio of 4.25:1). Projection I assumes the income elasticities of both grain and red meat are zero, that is, per capita consumption levels remain at their 1981 levels. Projection II assumes the income elasticity of grain is 0.22 and of meat 0.37. Projection III assumes the income elasticity of grain is 0.37 and of meat 1.34. See the Appendix for a discussion of income elasticities.

[c]Estimated net trade and stock changes is equal to domestic grain production minus estimated grain disappearance.

[d]No data is available for actual net stock changes during 1978-81.

129

decentralization of foreign exchange management in the late seventies. Pragmatic import and export targets for the Sixth Five-Year Plan released in 1982 called for continued balance of payments deficits through 1985. Beginning in 1980, however, the government implemented certain measures to regain a trade balance, and as a result achieved balanced trade in 1981 and a surplus in 1982. (See Table 9.) These measures included the cancellation or postponement of several large import contracts and tightened control of foreign exchange to reduce imports, as well as the establishment of special export zones and tax advantages to promote exports.

Table 9

Foreign Trade
(billion current yuan)

	I Value of Exports	II Value of Imports	III Net Exports (column I-column II)
1952	2.71	3.75	-1.04
1957	5.45	5.00	0.45
1965	6.31	5.53	0.78
1971	6.85	5.24	1.61
1972	8.29	6.40	1.89
1973	11.69	10.36	1.33
1974	13.94	15.28	-1.34
1975	14.30	14.74	-0.44
1976	13.48	12.93	0.55
1977	13.97	13.28	0.69
1978	16.77	18.74	-1.97
1979	21.17	24.29	-3.12
1980	27.24	29.14	-1.90
1981	36.76	36.77	-0.01
1982	41.43	35.77	5.66
1985 plan	(40.2)	(45.3)	(-5.1)

Sources: 1952-80: *Zhongguo Tongji Nianjian, 1981*, p. 353.
1981: *Zhongguo Jingji Nianjian, 1982*, p. VIII-32.
1982: *Beijing Review* 26 (19), p. IX.
1985: *Beijing Review* 25 (51), p. 17. These are Sixth Five-Year Plan targets.

Throughout the late seventies and early eighties China's net grain imports have shown an uninterrupted and substantial rise: the quantity of net grain imports more than doubled from 5.7 million tons in 1977 to 11.8 million tons in 1980, and by 1982 grew another 27% over 1980 to reach 15 million tons. These increases reflect continued rapid expansion in grain imports accompanied by slight reductions in grain exports. (See Table 10.) Expanded grain imports have been used to help make up shortfalls between domestic grain production

Table 10

Imports and Exports of Grain[a]
(1,000 m.t. trade grain)

	I Imports	II Exports	III Net Imports[d]
1952	0.1	1,528.8	-1,528.7
1957	166.8	2,092.6	-1,925.8
1965	6,405.2	2,416.5	3,988.7
1971	3,173.2	2,617.5	555.7
1972	4,756.2	2,925.6	1,830.6
1973	8,127.9	3,893.1	4,234.8
1974	8,121.3	3,643.9	4,477.4
1975	3,735.0	2,806.1	928.9
1976	2,366.5	1,764.7	601.8
1977	7,344.8	1,657.0	5,687.8
1978	8,832.5	1,877.2	6,955.3
1979	12,355.3	1,650.8	10,704.5
1980	13,429.3	1,618.3	11,811.0
1981	15,377[b]	1,260.8[b]	14,116.2
1982	15,546[c]	500[c]	15,046

Sources: [a]All figures from *Zhongguo Jingji Nianjian, 1982,* p. VIII-47, VIII-59, unless otherwise noted. Imports have been mainly wheat, which accounted for between 70% and 94% of total grain imports from 1977 through 1980. Exports have been primarily rice, which accounted for 50% to 90% of total exports during these years.

[b]World Bank, *China: Recent Economic Trends and Policy Development,* p. 138.
[c]USDA, *China: World Agriculture Regional Supplement,* 1983, p. 28.
[d]Column I minus Column II.

and consumption. Imports have been mainly wheat, but include some corn and soybeans, and they have been used primarily to supply urban areas.

All of China's grain imports are secured under bilateral government agreements. Most have been of three to four year duration, although a one-year contract was signed with Thailand in 1982. All existing grain import agreements but one (as of 10/26/83) are scheduled to terminate in 1984. The expiring commitments specified total imports of about 13 million tons annually. The one exception, an agreement to import between 3.5 and 4.2 million tons of wheat from Canada annually, will expire in mid-1985.

Although the future levels of grain import contracts are yet to be negotiated, China will probably continue to rely on grain imports to offset future imbalances between domestic grain production and consumption. To close the gap completely between projected production and disappearance discussed above, by 1985 net grain imports will probably have to more than double and by 1990 more than triple from their 1981/82 levels (see Table 8 columns CII and CIII). Unless the international price of grain drops considerably, by 1985 and 1990 the foreign exchange costs of meeting grain deficits solely through importing grain could therefore be substantial.

If exports continue to grow faster than grain imports, as they did during the late 1970s, then these costs could be offset by increased foreign exchange earnings. As of 1982, grain imports only accounted for about 10% of the value of merchandise exports. Merchandise exports would have to grow at an average annual rate of 15% or more in real terms through 1990 for this percentage not to rise. The future performance of China's export sector is unclear. Between 1978 and 1982 exports grew an average 25% a year, but the growth declined to 13% in 1982, and there was reportedly no growth in exports in 1983. If growth were 10%, then by 1990 the cost of grain imports would reach about 15% of merchandise exports' value.

Isolated Soy Protein: An Alternative to Feedgrain Imports

An alternative to importing grain to close the gap between domestic production and demand is importing isolated soy protein (ISP) as a protein ingredient in meat products. Imported ISP could substitute for some of the increase in domestic meat production and so mitigate rising feedgrain demand. One kilogram ISP can be used to produce a combination equivalent to 4 kilograms boneless lean meat, which is equivalent to 5.71 kilograms meat carcass. Moreover, production of one kilogram meat carcass requires 4.25 kilograms feedgrain. Therefore if the Chinese government chose to import one kilogram ISP, it could save 5.71 kilograms meat carcass times 4.25 kilograms feedgrain, or a total of 24.3 kilograms feedgrain.

According to industry sources, the 1983 c.i.f. Shanghai price of ISP in powder form was $3.00 per kilogram. Therefore, if the cost of importing 24.3 kilograms of grain exceeded $3.00, the Chinese government would save foreign exchange by importing ISP powder instead of importing feedgrain. In other words, if the border price of grain exceeded $124 a ton, then importing ISP powder instead of feedgrain would save the government foreign exchange.[18] In addition, since the cost of mixing ISP products with meat is probably less than the costs of nongrain inputs into meat production—labor, capital, transportation, and refrigeration—using ISP might also save some domestic resource costs.

Data for the 1983/84 production year show f.o.b export prices for U.S. No. 3 yellow corn ranging from $144 to $174 a ton.[19] Ocean freight and insurance costs would probably add an extra $20-25 a ton, resulting in a c.i.f. corn price per ton falling between $164 and $200. These prices indicate that as of 1983/84, it would save the Chinese government foreign exchange to import ISP powder instead of corn for feedgrain. If, for example, the Chinese government wished to increase the supply of meat carcass by one million tons (about 13% of 1981 output, equivalent to 0.7 million tons lean, boneless meat), it could do so by raising more animals using imported corn feed. At a grain:meat conversion ratio of 4.25:1, the imported feed would cost U.S. $700 to $800 million. At a grain:meat ratio of 6:1, it would cost U.S. $984 to $1200 million. Alternatively, an equivalent increase in meat supplies could be accomplished by using 3 million tons of existing meat carcass output in combination with ISP powder and water. The required ISP powder would cost U.S. $525 million, a 25% to 54% saving over importing corn feed, depending on the marginal feedgrain:meat conversion ratio.

Given China's overloaded domestic transport network, the state would probably not import large quantities of corn feed. Corn feed would have to be transported to meat producers located in rural areas. Although the government is promoting integrated livestock-meat producing facilities close to urban centers, they are still in the developmental stage and will account for only a small proportion of total meat output. China's response to the grain imbalance, therefore, would probably be not to import corn, but to increase imports of wheat or rice to feed the urban population. This would permit reduction of rural grain procurement, allowing farmers to retain more grain for use as feed. It would also reduce the burden of grain shipment on the domestic transportation system. Since wheat and rice command higher prices than corn on international markets, the foreign exchange argument for ISP powder is reinforced.

Whether or not ISP powder imports will save foreign exchange in the future depends mainly on two factors: relative movements in future corn and ISP prices, and future trends in China's feedgrain:meat ratio. ISP imports' profitability would be enhanced if international grain prices rise more rapidly than ISP prices, or if the feedgrain required per ton of meat increases. Lower grain prices or improved efficiency in the livestock sector, however, would reduce the profitability of importing ISP instead of grain. On this latter point, it would be profitable for China to use ISP powder in combination with red meats like pork and beef but not poultry. Since poultry requires considerably less feedgrain per unit meat—about 2 kilograms grain for each kilogram poultry meat (0.7 kilograms boneless poultry meat)—importing ISP would save only about 11 times its weight in feedgrain. International feedgrain prices would thus have to exceed about $270/ton for imports of ISP powder as a poultry meat ingredient to save foreign exchange.

The above paragraphs demonstrate that importing ISP is economically attractive for China, but only in comparison to the alternative of importing grain to produce red meat. In comparison to certain other alternatives, ISP may not be so attractive. Poultry meat production requires less than half the feedgrain used to produce an equivalent quantity of red meat. It also requires only about two-thirds the feedgrain required to produce the red meat used in an equivalent quantity of ISP-red meat mixture (mixed at the recommended ratio of 16 parts red meat to 1 part ISP powder and 3 parts water). ISP therefore does not compare well to poultry meat production. Traditional soy products could possibly also prove more economical than the ISP-red meat mixture. Further research would be required to ascertain exactly how the two compare. If the Chinese government were willing to promote production and consumption of poultry and traditional soy foods to reduce red meat feedgrain requirements, even more grain might be saved than if ISP were imported. At present poultry consumption is only about one tenth that of red meat, so there is probably significant potential for increase. One important constraint on poultry consumption, however, is the lack of poultry processing facilities— consumption remains low partly because purchasers must slaughter, pluck, and prepare the birds by hand.

Aside from the foreign exchange question, ISP may appeal to Chinese planners because it could save scarce domestic resources used in transportation and storage. At present, domestic transportation, harbor facilities, and energy resources constitute important bottlenecks in China's economy. As mentioned above, one way of easing domestic transportation and energy bottlenecks is to let agricultural products remain in rural areas thus saving the space and energy required to transport them, and to supply urban areas with imports. China uses grain imports in exactly this manner. Transportation of meat incurs even greater domestic resource costs than grain as it either must be shipped and stored in refrigerated facilities, which are practically nonexistent in China, or inefficiently on the hoof. ISP imports could thus save domestic transportation and energy resources by reducing the amount of rural-urban meat trans-shipment. If substituted for grain imports, ISP imports could ease the strain on inadequate grain-handling facilities at China's ports. Finally, ISP's long shelf life requiring no refrigeration would also reduce the needed capacity for cold storage.

Domestic transportation bottlenecks, however, are a two-edged sword: on the one hand, they make ISP attractive for enhancing urban meat supplies. On the other hand, they effectively limit the profitable distribution of ISP to urban and coastal areas. Nevertheless, urban areas are probably the most appropriate market for ISP because urban residents enjoy higher incomes and consume more meat per capita than the rural population.

Although China produces canned meats and sausages, at the present time the meat-processing industry is probably not in a position to make effective use of ISP meat ingredients on a large scale. First, the amount of meat processed is probably small. No data are available for meat, but only 5 to 10% of China's total food output is processed as compared with about 90% in the United States.[20] Second, food processing for the most part takes place in decentralized, small-scale, nonstandardized plants. Distribution of ISP to the individual plants and quality control over its use would thus pose difficulties. The exception to this might be in the processing of food for export. As a potential earner of foreign exchange, food processing for export has recently been promoted by the government. China has signed agreements for joint development of food processing for export with foreign firms.[21]

Although China's food-processing industry is currently underdeveloped, the Chinese government has recognized its potential for growth and the potential economic benefits from its growth. Processing perishables such as meat, fruit, fish, and milk where they are produced will reduce spoilage and the need for special storage and transportation facilities. Moreover, state-owned food factories have been an important source of state revenues, accounting for 11% of total profits and tax turned over to the state by the country's entire industrial establishment.[22] Their expansion would contribute further to state budgetary revenues. For these reasons, the government has called for growth in food processing.[23] As China's food processing industries develop, the scope for use of ISP in processed meat products will expand.

Aside from the meat-processing industry, in the short run an appropriate market for ISP might be urban institutional feeding. Urban China is organized around "units" (dan wei), the institutions where people work—factories, schools, ministries, military installations, department stores, and so on. Units provide their employees not only with jobs, but often with housing, schools, health care, food, and other amenities, all on or near the unit's location. Many units in urban China have cafeterias or dining halls that serve regular meals, saving individuals the considerable time required to shop for and prepare meals on their own. For units with more than one branch and large numbers of employees living on or near the workplace, institutional feeding can take on large proportions. Institutional feeding may therefore be an appropriate arena for use of ISP. Further research would be necessary to determine the degree of centralization in institutional food management, how institutional feeding is managed, how institutions purchase food, and what prices they pay and charge for food in order to determine how best to gain access to and distribute ISP in the institutional food industry.

With some exceptions, Chinese food is cut or chopped into small pieces, cooked by deep-frying, stir-frying, boiling, stewing, or steaming, and served with rice (in the South) and wheat noodles or steamed bread (in the North). Meat is purchased in whole pieces and chopped or minced in the kitchen before it is cooked. In daily meals meat is used sparingly. It is often used for a meat-flavored sauce to serve over noodles or rice, as a noodle filling (in jiao zi or wonton), or to add flavor and texture to a vegetable or bean curd dish. Dishes in which meat is the major ingredient are served less frequently and reserved for special occasions. As incomes increase, however, such dishes will probably become more common.

The appropriate form for ISP would be powder, granules, shreds, chunks, or slices. ISP powder or granules mixed with ground or finely diced meat could be used in a wide range of popular dishes: stuffed noodles (jiao zi, wonton, bao zi), meat sauces over noodles (danzi mian, jajiang mian), and certain stir-fry dishes and soups. ISP in the form of meat chunks and slices could be used for stir-fry, soups, stews, or steamed dishes. From the state's point of view, the price of imported ISP in form other than powder, however, would have to be low enough to save foreign exchange.

Conclusion

Analysis of recent consumption and production trends indicates that if incomes, population, and production increase as planned, the demand for grain would probably exceed domestic supply by a growing margin through the 1980s. The imbalance between grain production and demand results primarily from increases in and the changing composition of food demand caused by rising personal incomes and population. Major factors determining the extent of the grain gap are: the rates of population and income growth, growth in grain production, and the efficiency of feedgrain use in the meat sector.

Some supply-side approaches to meeting the imbalance have been discussed above. One such approach is to maintain growth of grain output at rates exceeding the plan targets. Another is to improve the efficiency of meat production so as to reduce feedgrain requirements. A third option is to depend on imports of grain and ISP to close the gap. As discussed in the previous section, as a substitute for feedgrain for hogs or cattle, ISP imports would save the government foreign exchange. ISP is also more easily transported and stored than red meat. It could be used in the "middle agricultural sector" between animal production and the consumer. As a substitute for grain to feed poultry, however, ISP imports would not be economical. Since poultry production uses less than half the amount of grain required to produce red meat, substitution of poultry products for red meat would conserve more grain and more foreign exchange than ISP imports.

In addition to boosting grain availability through supply-side measures, the government could also implement policies to control demand. Reduced emphasis on consumption and personal incomes is one alternative. The government could take measures to shift a greater proportion of national resources from consumption to accumulation and to divert national income increases from private pockets into public coffers. Such measures would reduce growth in per capita incomes and thus reduce the demand for grain and meat.

Prices could also be used to help correct the imbalance. Higher retail sales prices for grain and especially for meat would reduce demand. Price hikes would undoubtedly be unpopular, and so for political reasons the government might wish to avoid them. The present subsidies to urban consumers resulting from low state retail prices, however, represent a substantial drain on the state budget and have contributed to recent budget deficits. The cost to the government incurred by not passing 1979 procurement price increases for key agricultural products along to urban consumers averaged 13 billion yuan or about 12% of state budgetary revenues annually during 1979-81. In 1982, the cost reached 17 billion yuan, more than 15% of state budgetary revenues.[24] During this three-year period, each urban resident received more than 150 yuan in state price subsidies for grain and oil consumed.[25] Despite political pressures to maintain low urban food prices, the state may not be able to continue such subsidies indefinitely. Retail price increases would improve both the grain imbalance and improve the state's financial situation.

Finally, the state could employ rationing to maintain the balance between grain supply and realized consumption. Rationing supplies of grain and meat to urban consumers would reduce per capita urban consumption regardless of income levels. Reducing the supply of grain to rural households would not only lower rural consumption of grain, but would also reduce feedgrain availability and thus divert grain from animal raising. Such quantity controls have been used extensively in the past and could easily be revived. Rationing, however, could have negative side effects. If ration levels are inappropriate, they can lead to inefficient allocation of resources. Moreover, rationing lowers the purchasing power of personal incomes, and so could reduce incentives to work.

Although the projections in this paper predict considerable growth in total grain demand, it is important to remember that given China's large population, even a small change in per capita consumption levels translates into a significant aggregate consumption increase. Thus a small error in the per capita projections for 1985 and 1990 would lead to a large error in the projections for total demand. The grain gap's projected magnitude, therefore, should be interpreted as a mean estimate with wide possible variance.

Postscript

Since the completion of the original paper, several developments have taken place in China's grain and livestock economy. These developments, as well as the publication of new data, make possible a re-examination of the original conclusions and projections.

A summary of recently released data on China's consumption and production of grain and meat are given in Table 11. Several conclusions can be drawn from these data. First, during the past few years, growth in per capita consumption and total grain disappearance have been consistent with the original predictions. For both grain and red meat, per capita consumption in 1984 is slightly lower than the "moderate" projections, and if growth continues at the same rates, actual per capita consumption in 1985 will fall well within the predicted ranges. Population has been growing more slowly than planned, but this is offset by an upward revision in the data for the total population count following publication of results from the 1982 census. These trends in per capita consumption and population suggest that total grain disappearance in the mid-1980s should fall within the predicted range given in Table 11.

Table 11

Updated Information on China's

Grain and Meat Economy

	1984 Actual[a]	1985 Predicted or Target Level[b]		
		Low	Moderate	High
Per Capita Grain Consumption (kg.)	230	219	231	239
Per Capita Red Meat Consumption (kg.)	13.3	11.8	14.0	16.0
Population (millions)	1,036		1,060	
Total Grain Disappearance (1,000 tons)	389,706	372,000	397,000	416,000
Grain Output (1,000 tons)	407,120		360,000	
Meat Output	15,250	14,000		14,600

[a]Taken from Sicular, 1985, and *Beijing Review* 28(15), March 25, 1985, p. VIII.

[b]Taken from Tables 3, 4, 6 and 8 above.

On the production side, actual performance has surpassed the national planning targets by a substantial margin. Red meat output grew at an average annual rate of 4.9% between 1981 and 1984, almost double the target rates of 2 to 2.6% a year. Grain output grew 5.8% a year over this period, also considerably faster than the target rate of 2.6% a year. As a result, by 1984 production of both meat and grain surpassed their 1985 target levels. Even if reports of no growth or even a decline in grain output for 1985 prove true, actual grain production will probably exceed its target by 40 to 50 million metric tons. This additional grain by far outweighs the unexpectedly high feedgrain requirements of the rapidly growing livestock sector.

Due to these increases in grain production, the actual gap between total grain demand and supply will be smaller than predicted for both 1985 and 1990. Rather than experiencing an overall grain deficit in 1985, China's grain situation will likely be in balance, or even show a small surplus.

The implications of these developments for the use of ISP in China are mixed. The argument for importing ISP as an alternative to importing feedgrains appears less compelling if China's import needs have been considerably reduced. Whether or not ISP imports save China foreign exchange, however, still depends on the relative prices of ISP and feedgrain in international markets. If the border price of grain exceeds $124 a ton, then importing ISP powder should continue to save foreign exchange. The simple cost-benefit calculation in the original paper holds regardless of whether China is a grain importer or an exporter. If China is an importer, then China would save foreign exchange by importing ISP instead of feedgrain. If China is a grain exporter, then importing ISP powder and exporting the feedgrain that is saved would similarly earn foreign exchange. With the low international grain prices in 1985, however, ISP remains only a marginally attractive alternative to China, a country short of foreign exchange.

Although supply and demand are balanced in the aggregate, certain imbalances across regions and in the composition of output remain. Regional imbalance exists because livestock production is most developed in south China, which has surplus rice but insufficient corn and soybeans. In contrast, China's North and Northeast product surplus corn and soybeans which are currently being exported (see Wiens, 1985). Such regional differences suggest that even if China's aggregate grain production is equal to total demand, in the short term feed grain imports will continue to be required in the South. In the longer run, as China's internal transport develops and as livestock production is shifted North, regional import demands may be reduced.

Imbalances in the composition of meat output also modify the picture. Oversupply of pork fat remains a problem, while demand for lean meat goes unsatisfied. Imports of ISP powder could help solve this disproportion of fat to lean. New price reforms implemented in 1985 raise the relative price of lean meat and may eliminate this imbalance in the longer run. (See Sicular, 1985.)

Despite spectacular growth in China's grain production, therefore, ISP powder imports may remain viable, at least in the short term. ISP imports remain attractive because of persistent regional and compositional imbalances within China. Ultimately, a careful, more detailed cost and benefit analysis of the type described in the chapter by F.H. Schwarz is required.

References

Sicular, Terry, "China's Grain and Meat Economy: Recent Developments and Implications for Trade," *American Journal of Agricultural Economics*, 67(5), December, 1985.

Wiens, Thomas B., "Discussion: Agriculture in the Soviet Union and China: Implications for Trade," *American Journal of Agricultural Economics*, 67(5), December, 1985.

FOOTNOTES

1. Bo Yi-bo, "Several Questions, on Developing the National Economy Proportionately and in a Planned Way," *Hongqi*, No. 19, Oct. 1, 1983, pp. 2-9. Translated in Foreign Broadcast Information Service (hereafter FBIS), *Daily Report*, 3 November 1983, pp. K2-K12.

2. Between 1978 and 1981 rural per capita income in nominal terms including noncollective income is estimated to have risen from 111.74 yuan to 194.53 yuan. Over these three years, the rural cost of living is estimated to have risen 23.3%. (See Appendix 1.) These figures imply an average annual growth rate in real income per capita of 12%. In urban areas, average wages (Table II) rose from 614 to 772 yuan, and the cost of living increased 12.3%. A survey of urban households shows that between 1978 and 1981 the dependency ratio declined from 2.06 to 1.77 *(Beijing Review*, 26 April 1982, p. 16). These figures imply an average annual growth rate in real income per capita urban resident of 9%.

3. Income and expenditure increases are in nominal terms. Rural household survey data is taken from *Zhongguo Tongji Nianjian, 1981*, pp. 431-433. Urban household survey data is taken from W. Klatt, "The Staff of Life: Living Standards in China, 1977-81," *China Quarterly*, No. 93, March 1983, p. 41.

4. In 1981 an agricultural household survey showed per capita poultry consumption equal to 0.71 kilograms for rural households, and an urban household survey showed per capita poultry purchases at 1.92 kilograms. Weighting each figure by its respective share of the population (the rural population is 80% of the total), I calculate national average poultry consumption of 0.95 kilograms per capita. 1981 national average fresh egg consumption was 2.44 kilograms per capita. See *Zhongguo Tongji Nianjian, 1981*, pp. 430, 434, 439. The per capita milk consumption figure of 1.4 kilograms is given in U.S.D.A. Economic Research Service, *China: World Agricultural Regional Supplement, Review of 1982 and Outlook for 1983*, 1983, p.15.

5. In other developing Asian countries income elasticities of demand for grain range from 0.15 to 0.3 and those for meat from 0.55 to 1.15. See J. Kilpatrick, "China: The Drive for Dietary Improvement," in Joint Economic Committee of the U.S. Congress, *China Under the Four Modernizations* (U.S. Government Printing House, Washington, D.C. 1983), p. 455.

6. For urban areas, state retail sales prices may understate the increases in prices consumers pay for grain and meat. These commodities are also purchased at free market and state negotiated prices, which are somewhat higher. For rural consumers, the state above-quota procurement prices for grain and hogs represent the opportunity cost of retaining these items for own consumption. State above-quota procurement prices are therefore used to calculate increases in consumer prices for grain and meat.

7. This estimate assumes 4.25 kilograms grain are required to produce one kilogram red meat. Throughout this paper red meat feed conversion is stated in terms of quantity of grain required per quantity meat carcass rather than per quantity live weight. Due to the incomplete nature of Chinese data, use of the meat carcass conversion ratio requires fewer assumptions and is more convenient. According to industry sources, meat carcass is generally equivalent to about 70% of live weight. The 4.25:1 carcass weight conversion ratio is therefore approximately equivalent to a 3:1 live-weight conversion ratio.

8. *Zhongguo Nongye Nianjian, 1981*, p. 109, and *Renmin Ribao*, 7 January 1982, p. 3, translated in FBIS, *Daily Report*, 18 January 1982, p. K12.

9. FBIS, *Daily Report*, 17 May 1983, p. K18.

10. FBIS, *Daily Report*, 11 January 1984, p. K22.

11. According to the Chinese definition, grain includes legumes such as soybeans, mung and azuki beans, field peas, and broad beans, as well as the usual cereals such as rice and wheat.

12. FBIS, *Daily Report*, 3 January 1984, p. K7.

13. Feedgrain demand (D^{fg}) = Red Meat Demand (D^m) x Feedgrain Required per Unit Meat (α); therefore,

$$\frac{\delta D^{fg}/\delta t}{D^{fg}} = \frac{\delta D^m/\delta t}{D^m} + \frac{\delta a/\delta t}{\alpha}$$

14. World Bank, *China: Recent Economic Trends and Policy Developments*, March 31, 1983, p. 9.

15. Census statistics indicate a 1.46% rate of natural increase in 1981 (FBIS, *Daily Report*, 28 October 1982, p. K1, and John S. Aird, "China's Surprising Census Results," *China Business Review*, 10(2), March-April 1983, pp. 10-15). Sample survey data published by the State Statistical Bureau shows a natural growth rate of 1.45% (FBIS, *Daily Report*, 2 May 1983, p. K16). At a rate of 1.46%, by 1990 China's population would reach 1,135 million, a number not significantly different from the projected population at planned rates. (See Table 4.)

16. See *Guangming Ribao*, June 7, 1980, p. 4, and *Zhongguo Nongye Nianjian, 1981*, p. 426.

17. For example, see *Zhongguo Nongye Nianjian, 1981*, pp. 425-428, FBIS, *Daily Report*, August 4, 1982, p. K14-K15, and *Guangming Ribao*, June 7, 1980, p. 4.

18. In general, substitution of ISP imports substituted for feedgrain imports saves foreign exchange when the following inequality holds:

$$p^{fg} > p^{isp} \times \frac{b}{\alpha h}$$

   where

   p^{fg} = c.i.f. price of feedgrain

   p^{isp} = c.i.f. price of ISP

   h= the amount of extended meat produced per unit of ISP

   b = the amount of boneless meat per unit meat carcass

   α = feedgrain required to produce one unit meat carcass

19. USDA, *Foreign Agricultural Circular: Grains*, 11/15/83.

20. Kenneth I. Bowman, "Food Processing," *China Business Review*, 10(2), March-April 1983, p. 6.

21. Ibid, pp. 6-9.

22. FBIS, *Daily Report*, 17 June 1982, p. K19-20, and 8 July 1983, p. K19-20.

23. FBIS, *Daily Report*, 17 June 1982, pp. K19-20 and 8 July 1983, p. K19-20.

24. Martin Weil, "The Sixth Five-Year Plan," *China Business Review*, 10(2), March-April 1983, pp. 22-23.

25. FBIS, *Daily Report*, 16 March 1982, p. K11.

Appendix 1: Estimation of Income Elasticities for Grain and Red Meat Demand

In China per capita consumption q_i of a good is the function of prices p, quantity availabilities to consumers or quota rations r, and income Y:

$$q_i = q_i \ (p,r; Y).$$

Observed changes in per capita consumption therefore reflect consumer responses to changes not only in income, but also in prices and quotas.

Observed percentage increases in per capita consumption are thus the sum of percentage increases in prices, quotas and income each multiplied by its respective elasticity of consumption:

$$\frac{dq_i}{q_i} = \sum_j \epsilon_{ij} \frac{dp_j}{p_j} + \sum_j \rho_{ij} \frac{dr_j}{r_j} + \eta_i \frac{dY}{Y}$$

where P_j = the price of good j

ϵ_{ij} = the elasticity of good i with respect to the jth quota

r_j = the quota ration for good j

ρ_{ij} = the elasticity of good i with respect to the jth quota

Y = income

η_i = the income elasticity of good i

In order to estimate the income elasticity of a consumer good, one should separate out the effects of price and quota changes on consumption. This, however, requires data on per capita consumption, prices, quotas and income. Unfortunately, the data available for China are inadequate to separate out price, quota and income effects.

Due to this absence of data, the income elasticities used in this paper are only rough approximations of the relationships between income and consumption of grain and meat. The elasticities are simply the percentage change in national average per capita consumption divided by the percentage change in per capita income. The estimated income elasticities $\hat{\eta}_i$ therefore contain price and quota effects. The relationship between the estimated and the true income elasticities η_i are

$$\hat{\eta} = \frac{dq_i/q_i}{dY/Y} = \sum_j \epsilon_{ij} \frac{dp_j/p_j}{dY/Y} + \sum_j \rho_{ij} \frac{dr_j/r_j}{dY/Y} + \eta_i \ .$$

If prices and quotas have changed and if ϵ_{ij} and ρ_{ij} differ from zero, then $\hat{\eta}_i$ may be a biased estimate of η_i.

In fact, between 1978 and 1981 prices and quotas did change. In urban areas,

143

the state retail sales price of grains remained constant, and those of red meats increased about 30%. (Pork prices rose 32%, and beef and mutton prices 33%. See W. Klatt, "The Staff of Life: Living Standards in China 1977-81," *China Quarterly*, March 1983, No. 93, p. 42.) Concurrently, meat rationing was suspended in most cities, and the general availability of grain and meat improved. In rural areas, the prices farmers received for sale of grain and hogs to the state and in free markets rose, and so the opportunity cost of retaining those products for consumption increased. State above-quota procurement prices for grain rose 45% and for red meat 28%. (See *Zhongguo Tongji Nianjian, 1981*, p. 405.) The availability of grains and meats to rural consumers also increased as the state began to sell more of these products in rural areas and as free markets expanded.

How did these price and quota changes affect consumption? On the one hand, price increases probably reduced levels of grain and meat consumption. An income elasticity not corrected for such price increases would therefore be too low. On the other hand, easing of quotas and increased availabilities probably had a positive impact. An income elasticity not corrected for such quota adjustments would therefore be too high. If the price and quota effects exactly offset each other, then the estimated income elasticities $\hat{\eta}_i$ would equal the true income elasticities η_i. If the price effects outweighed the quota effects, then the estimated income elasticities would understate the true income elasticities. Conversely, if the quota effects outweighed the price effects, then the estimated elasticities would overstate the true elasticities.

Even with the above simplifications, calculation of crude income elasticities is not straightforward. Although information on per capita incomes and consumption are available, they are not ideally suited to the task. Certain assumptions and patching together of available information are required in order to calculate the percentage changes in consumption and income. The steps followed in calculation of the income elasticities are presented below.

Step One: Calculation of Changes in Per Capita Consumption

The first step in the estimation of income elasticities for grain and meat is to determine the change in consumption levels between 1978 and 1981. Published Chinese statistics on consumption are sparse. Table III presents much of the available data.

National average consumption data for 1978 and 1981 are available for grain and pork. These data show per capita grain consumption rose 12% and pork consumption 45% over the three-year period. Under the assumption that the proportion of pork in red meat consumption remains constant, per capita red meat consumption increased by the same percentage as pork.

Unfortunately, consumption data disaggregated by region or sector are insufficient for calculation of meaningful and comparable disaggregated income elasticities. Therefore only national average income elasticities can be estimated.

Step Two: Calculation of Changes in Per Capita Incomes, Unadjusted for Inflation

China does not publish data for national average personal income per capita, and available income data are fragmentary. Estimation of the percentage increase in personal income per capita therefore requires patching together different pieces of information. As a result, the absolute levels of personal income presented below are probably questionable. It is not unreasonable to assume, however, that growth in these calculated income levels paralleled growth in the true levels between 1978 and 1981.

Calculation of growth in per capita incomes uses data on rural and urban incomes. With respect to rural incomes, national average per capita rural income distributed by collectives was 74.00 yuan in 1978 and 101.32 yuan in 1981 (see Table II). No data is available on national average per capita income from noncollective sources, that is, from household private sidelines and other sources. A survey of rural households, however, gives sample average noncollective income per capita (*Zhongguo Tongji Nianjian, 1981*, p. 431). The survey data shows sideline and other income was equal to 51% of sample average collectively distributed income in 1978, and 92% in 1981. Assuming that this ratio between collective and noncollective incomes is representative, national average noncollective income per capita would equal 37.74 yuan in 1978 and 93.21 yuan in 1981. National average per capita rural incomes would therefore equal 111.74 yuan in 1978 and 194.53 yuan in 1981.

Total rural personal income is per capita income multiplied by a rural population figure. Two alternative population figures are used.

	Members of Collectives[a] (millions)	Total Rural Personal Income I (million yuan)	Rural Population[b] (millions)	Total Rural Personal Income II (million yuan)
1978	803.20	89.750	838.15	93,655
1981	818.81	159,283	857.52	166,813

Sources: a. *Zhongguo Tongjji Nianjian, 1981*, p. 131.
   b. Ibid, p. 89.

Total national personal income is calculated as the sum of rural personal incomes above and the total wage bill of urban collective and state enterprises.

	Total Urban Collective and State Enterprise Wage Bill*	Total Rural Personal Income: Estimate I	Estimate II	Total National Personal Income: Estimate I	Estimate II
1978	56,920	89,750	93,655	146,670	150,575
1981	82,040	159,283	166,813	241,323	248,853

*Zhongguo Jingji Nianjian, 1982, p. VIII-7.

National per capita income is total personal income from above divided by population. For estimate I, the divisor is urban population plus the member population of rural collectives. For estimate II, the divisor is urban population plus total rural population. (Urban population figures are taken from *Zhongguo Tongji Nianjian, 1981*, p. 89.)

	Estimate I			Estimate II		
	Total Personal Income (million yuan)	Rural Collective Plus Urban Population (millions)	National Average Per Capita Income (yuan)	Total Personal Income (million yuan)	Rural Plus Urban Population (millions)	National Average Per Capita Income (yuan)
1978	146,670	923.14	159	150,575	958.04	157
1981	241,323	957.51	252	248,853	996.22	250
1981 / 1978	1.65	1.04	1.58	1.65	1.04	1.59

The percentage increase in national average per capita incomes between 1978 and 1981 comes out at 58-59%. The average annual increase was 16.6%.

Step Three: Calculation of Changes in Per Capita Incomes Adjusted for Inflation

The per capita income figures above are in nominal terms. Since the price level rose between 1978 and 1981, the percentage change in real income was less than 58-59%. In order to calculate the change in real incomes, the 1981 figures should be deflated, that is, converted into 1978 currency equivalents.

A number of price indexes are published by China's State Statistical Bureau. Little is known about how they are calculated, so it is difficult to ascertain whether they are biased. Moreover, it is not always clear which index should be used to deflate which type of income. In the absence of better alternatives, the available price indexes must suffice.

Urban income is deflated using the cost of living index for workers. This index shows prices rose 12.3% between 1978 and 1981. It reportedly covers not only state retail sales prices, but also negotiated and free market prices. For rural income, two price indexes seem relevant. The agricultural procurement price index (which includes quota, above-quota, and negotiated procurement prices but, unfortunately, not rural free market prices for agricultural products) measures the opportunity cost to farmers of retaining products for own consumption. This index therefore gives the price increase for products to farmers, and it is used to deflate income spent on agricultural products. It shows prices rose 38.5% between 1978 and 1981. For lack of a better alternative, the national retail sales price index is used to deflate farm income not spent on agricultural products. This index shows prices increased 10.7% between 1978 and 1981. The national retail sales price index covers state retail sales prices, negotiated sales prices, and market prices. (All price indexes are taken from *Zhongguo Tongji Nianjian, 1981*, p. 403.)

In deflating rural income, it is assumed that the proportion of income spent on agricultural products equals the proportion of income spent on foodstuffs given in the agricultural household survey data (*Zhongguo Tongji Nianjian, 1981*, p. 431-432). That proportion was 59% in 1978 and 51% in 1981. The remainder of rural income is deflated using the national retail sales price index. Deflated urban and rural incomes in 1978 yuan are given below.

	Urban Collective and State Enterprise Wage Bill (million 1978 yuan)	Total Rural Personal Income (million 1978 yuan)		Total National Personal Income (sum of urban and rural)	
		I	II	I	II
1978	56,920	89,750	93,655	146,670	150,575
1981	73,054	129,158	135,263	202,212	208,317
Factor of Deflation	112.3	123.3	123.3	119.3	119.5

Per capita real incomes are the deflated total income figures divided by population (see above for population figures).

National Average Per Capita Income
(constant 1978 yuan)

	I	II
1978	159	157
1981	211	209
$\frac{1981}{1978}$	1.33	1.33
Factor of Deflation	119.4	119.6

In real terms, per capita incomes increased 33% between 1978 and 1981 as compared to 58-59% for nominal incomes. The average annual rate of increase in real per capita incomes was 10.0%. Prices rose about 19% over this period. Inflation was higher for the rural than for the urban population.

Step Four: Calculation of Income Elasticities

Income elasticities are calculated as the average annual percentage changes in per capita consumption of grain and meat divided by the average annual percentage change in per capita incomes calculated above.

	Calculated Using Nominal Income	Calculated Using Real Income
For Grains	0.22	0.37
For Red Meat	0.81	1.34

Appendix 2: Projection of Total Grain Disappearance and Net Trade and Stock Changes under Alternate Assumptions

This appendix presents projections of total grain disappearance and net trade and stock changes using alternate assumptions with respect to per capita income growth, the income elasticities of demand for grain ($\hat{\eta}_g$) and meat ($\hat{\eta}_m$), the feedgrain conversion ratio, and grain production growth between 1981 and 1990. The alternate assumptions are: per capita income growth of 4%, 6%, and 8%; income elasticity pairs of $\hat{\eta}_g = \hat{\eta}_m = 0$, $\hat{\eta}_g = 0.22$ with $\hat{\eta}_m = 0.81$, and $\hat{\eta}_g = 0.37$ with $\hat{\eta}_m = 1.34$; feedgrain conversion ratios of 4.25 and 6.0; and grain production growth rates of 2% and 4%.

The numbers for total grain disappearance in this appendix may differ slightly from those in Tables 1-10 due to rounding. Note also that the net trade and stock changes given in this appendix are calculated using 2% and 4% 1981-1990 average annual growth rates for grain production rather than the 1985 and 1990 grain production targets used in the tables.

Projected Total Grain Disappearance and Net Trade and Stock Changes (1,000 mt. original grain)

Assumptions

		Projections			
		Total Grain		Net Trade and	
Per Capita Income Growth	4	Disappearance		Stock Changes	
Income Elasticity, Grain	0	1985	1990	1985	1990
Income Elasticity, Meat	0	372477	402879	-20665	-14450
Grain/Meat Conversion Ratio	4.25				
Grain Production Growth	2				

Assumptions

		Projections			
		Total Grain		Net Trade and	
Per Capita Income Growth	4	Disappearance		Stock Changes	
Income Elasticity, Grain	0	1985	1990	1985	1990
Income Elasticity, Meat	0	372477	402879	7750	59726
Grain/Meat Conversion Ratio	4.25				
Grain Production Growth	4				

Assumptions

		Projections			
		Total Grain		Net Trade and	
Per Capita Income Growth	4	Disappearance		Stock Changes	
Income Elasticity, Grain	0	1985	1990	1985	1990
Income Elasticity, Meat	0	394366	426234	-42554	-37805
Grain/Meat Conversion Ratio	6				
Grain Production Growth	2				

Assumptions

		Projections			
		Total Grain		Net Trade and	
Per Capita Income Growth	4	Disappearance		Stock Changes	
Income Elasticity, Grain	0	1985	1990	1985	1990
Income Elasticity, Meat	0	394366	426234	-14139	36371
Grain/Meat Conversion Ratio	6				
Grain Production Growth	4				

Assumptions

		Projections			
Per Capita Income Growth	4				
Income Elasticity, Grain	.22	Total Grain		Net Trade and	
Income Elasticity, Meat	.81	Disappearance		Stock Changes	
Grain/Meat Conversion Ratio	4.25	1985	1990	1985	1990
Grain Production Growth	2	389694	446238	-37882	-57810

Assumptions

		Projections			
Per Capita Income Growth	4				
Income Elasticity, Grain	.22	Total Grain		Net Trade and	
Income Elasticity, Meat	.81	Disappearance		Stock Changes	
Grain/Meat Conversion Ratio	4.25	1985	1990	1985	1990
Grain Production Growth	4	389694	446238	-9466	16366

Assumptions

		Projections			
Per Capita Income Growth	4				
Income Elasticity, Grain	.22	Total Grain		Net Trade and	
Income Elasticity, Meat	.81	Disappearance		Stock Changes	
Grain/Meat Conversion Ratio	6	1985	1990	1985	1990
Grain Production Growth	2	414560	477357	-62748	-88928

Assumptions

		Projections			
Per Capita Income Growth	4				
Income Elasticity, Grain	.22	Total Grain		Net Trade and	
Income Elasticity, Meat	.81	Disappearance		Stock Changes	
Grain/Meat Conversion Ratio	6	1985	1990	1985	1990
Grain Production Growth	4	414560	477357	-34333	-14752

Assumptions

Projections

Per Capita Income Growth	4				
Income Elasticity, Grain	.37	Total Grain		Net Trade and	
Income Elasticity, Meat	1.34	Disappearance		Stock Changes	
Grain/Meat Conversion Ratio	4.25	1985	1990	1985	1990
Grain Production Growth	2	401768	479127	-49956	-90698

Assumptions

Projections

Per Capita Income Growth	4				
Income Elasticity, Grain	.37	Total Grain		Net Trade and	
Income Elasticity, Meat	1.34	Disappearance		Stock Changes	
Grain/Meat Conversion Ratio	4.25	1985	1990	1985	1990
Grain Production Growth	4	401768	479127	-21541	-16522

Assumptions

Projections

Per Capita Income Growth	4				
Income Elasticity, Grain	.37	Total Grain		Net Trade and	
Income Elasticity, Meat	1.34	Disappearance		Stock Changes	
Grain/Meat Conversion Ratio	6	1985	1990	1985	1990
Grain Production Growth	2	428741	516492	-76929	-128063

Assumptions

Projections

Per Capita Income Growth	4				
Income Elasticity, Grain	.37	Total Grain		Net Trade and	
Income Elasticity, Meat	1.34	Disappearance		Stock Changes	
Grain/Meat Conversion Ratio	6	1985	1990	1985	1990
Grain Production Growth	4	428741	516492	-48514	-53887

Assumptions

		Projections			
Per Capita Income Growth	6				
Income Elasticity, Grain	0	Total Grain		Net Trade and	
Income Elasticity, Meat	0	Disappearance		Stock Changes	
Grain/Meat Conversion Ratio	4.25	1985	1990	1985	1990
Grain Production Growth	2	372477	402879	-20665	-14450

Assumptions

		Projections			
Per Capita Income Growth	6				
Income Elasticity, Grain	0	Total Grain		Net Trade and	
Income Elasticity, Meat	0	Disappearance		Stock Changes	
Grain/Meat Conversion Ratio	4.25	1985	1990	1985	1990
Grain Production Growth	4	372477	402879	7750	59726

Assumptions

		Projections			
Per Capita Income Growth	6				
Income Elasticity, Grain	0	Total Grain		Net Trade and	
Income Elasticity, Meat	0	Disappearance		Stock Changes	
Grain/Meat Conversion Ratio	6	1985	1990	1985	1990
Grain Production Growth	2	394366	426234	-42554	-37805

Assumptions

		Projections			
Per Capita Income Growth	6				
Income Elasticity, Grain	0	Total Grain		Net Trade and	
Income Elasticity, Meat	0	Disappearance		Stock Changes	
Grain/Meat Conversion Ratio	6	1985	1990	1985	1990
Grain Production Growth	4	394366	426234	-14139	36371

Assumptions

			Projections			
Per Capita Income Growth	6					
Income Elasticity, Grain	.22		Total Grain		Net Trade and	
Income Elasticity, Meat	.81		Disappearance		Stock Changes	
Grain/Meat Conversion Ratio	4.25		1985	1990	1985	1990
Grain Production Growth	2		398666	470517	-46854	-82088

Assumptions

			Projections			
Per Capita Income Growth	6					
Income Elasticity, Grain	.22		Total Grain		Net Trade and	
Income Elasticity, Meat	.81		Disappearance		Stock Changes	
Grain/Meat Conversion Ratio	4.25		1985	1990	1985	1990
Grain Production Growth	4		398666	470517	-18438	-7912

Assumptions

			Projections			
Per Capita Income Growth	6					
Income Elasticity, Grain	.22		Total Grain		Net Trade and	
Income Elasticity, Meat	.81		Disappearance		Stock Changes	
Grain/Meat Conversion Ratio	6		1985	1990	1985	1990
Grain Production Growth	2		425130	506316	-73318	-117887

Assumptions

			Projections			
Per Capita Income Growth	6					
Income Elasticity, Grain	.22		Total Grain		Net Trade and	
Income Elasticity, Meat	.81		Disappearance		Stock Changes	
Grain/Meat Conversion Ratio	6		1985	1990	1985	1990
Grain Production Growth	4		425130	506316	-44903	-43711

Assumptions

		Projections			
Per Capita Income Growth	6				
Income Elasticity, Grain	.37	Total Grain		Net Trade and	
Income Elasticity, Meat	1.34	Disappearance		Stock Changes	
Grain/Meat Conversion Ratio	4.25	1985	1990	1985	1990
Grain Production Growth	2	417447	525182	-65635	-136754

Assumptions

		Projections			
Per Capita Income Growth	6				
Income Elasticity, Grain	.37	Total Grain		Net Trade and	
Income Elasticity, Meat	1.34	Disappearance		Stock Changes	
Grain/Meat Conversion Ratio	4.25	1985	1990	1985	1990
Grain Production Growth	4	417447	525182	-37219	-62578

Assumptions

		Projections			
Per Capita Income Growth	6				
Income Elasticity, Grain	.37	Total Grain		Net Trade and	
Income Elasticity, Meat	1.34	Disappearance		Stock Changes	
Grain/Meat Conversion Ratio	6	1985	1990	1985	1990
Grain Production Growth	2	447270	572025	-95458	-183597

Assumptions

		Projections			
Per Capita Income Growth	6				
Income Elasticity, Grain	.37	Total Grain		Net Trade and	
Income Elasticity, Meat	1.34	Disappearance		Stock Changes	
Grain/Meat Conversion Ratio	6	1985	1990	1985	1990
Grain Production Growth	4	447270	572025	-67043	-109421

Assumptions

Projections

Per Capita Income Growth	8				
Income Elasticity, Grain	0	Total Grain		Net Trade and	
Income Elasticity, Meat	0	Disappearance		Stock Changes	
Grain/Meat Conversion Ratio	4.25	1985	1990	1985	1990
Grain Production Growth	2	372477	402879	-20665	-14450

Assumptions

Projections

Per Capita Income Growth	8				
Income Elasticity, Grain	0	Total Grain		Net Trade and	
Income Elasticity, Meat	0	Disappearance		Stock Changes	
Grain/Meat Conversion Ratio	4.25	1985	1990	1985	1990
Grain Production Growth	4	372477	402879	7750	59726

Assumptions

Projections

Per Capita Income Growth	8				
Income Elasticity, Grain	0	Total Grain		Net Trade and	
Income Elasticity, Meat	0	Disappearance		Stock Changes	
Grain/Meat Conversion Ratio	6	1985	1990	1985	1990
Grain Production Growth	2	394366	426234	-42554	-37805

Assumptions

Projections

Per Capita Income Growth	8				
Income Elasticity, Grain	0	Total Grain		Net Trade and	
Income Elasticity, Meat	0	Disappearance		Stock Changes	
Grain/Meat Conversion Ratio	6	1985	1990	1985	1990
Grain Production Growth	4	394366	426234	-14139	36371

Assumptions

		Projections			
		Total Grain		Net Trade and	
		Disappearance		Stock Changes	
		1985	1990	1985	1990
Per Capita Income Growth	8				
Income Elasticity, Grain	.22				
Income Elasticity, Meat	.81				
Grain/Meat Conversion Ratio	4.25				
Grain Production Growth	2	407888	496747	-56076	-108318

Assumptions

		Projections			
		Total Grain		Net Trade and	
		Disappearance		Stock Changes	
		1985	1990	1985	1990
Per Capita Income Growth	8				
Income Elasticity, Grain	.22				
Income Elasticity, Meat	.81				
Grain/Meat Conversion Ratio	4.25				
Grain Production Growth	4	407888	496747	-27661	-34142

Assumptions

		Projections			
		Total Grain		Net Trade and	
		Disappearance		Stock Changes	
		1985	1990	1985	1990
Per Capita Income Growth	8				
Income Elasticity, Grain	.22				
Income Elasticity, Meat	.81				
Grain/Meat Conversion Ratio	6				
Grain Production Growth	2	436027	537843	-84215	-149414

Assumptions

		Projections			
		Total Grain		Net Trade and	
		Disappearance		Stock Changes	
		1985	1990	1985	1990
Per Capita Income Growth	8				
Income Elasticity, Grain	.22				
Income Elasticity, Meat	.81				
Grain/Meat Conversion Ratio	6				
Grain Production Growth	4	436027	537843	-55799	-75238

Assumptions

		Projections			
Per Capita Income Growth	8				
Income Elasticity, Grain	.37	Total Grain		Net Trade and	
Income Elasticity, Meat	1.34	Disappearance		Stock Changes	
Grain/Meat Conversion Ratio	4.25	1985	1990	1985	1990
Grain Production Growth	2	433852	577663	-82040	-189234

Assumptions

		Projections			
Per Capita Income Growth	8				
Income Elasticity, Grain	.37	Total Grain		Net Trade and	
Income Elasticity, Meat	1.34	Disappearance		Stock Changes	
Grain/Meat Conversion Ratio	4.25	1985	1990	1985	1990
Grain Production Growth	4	433852	577663	-53625	-115058

Assumptions

		Projections			
Per Capita Income Growth	8				
Income Elasticity, Grain	.37	Total Grain		Net Trade and	
Income Elasticity, Meat	1.34	Disappearance		Stock Changes	
Grain/Meat Conversion Ratio	6	1985	1990	1985	1990
Grain Production Growth	2	466747	636063	-114935	-247635

Assumptions

		Projections			
Per Capita Income Growth	8				
Income Elasticity, Grain	.37	Total Grain		Net Trade and	
Income Elasticity, Meat	1.34	Disappearance		Stock Changes	
Grain/Meat Conversion Ratio	6	1985	1990	1985	1990
Grain Production Growth	4	466747	636063	-86520	-173459

Chapter 5

The Potential Role of Isolated Soy Protein Food Ingredients in Mexico

*G. Edward Schuh, Matthew G. Smith
and Maria Ignez Schuh**

*G. Edward Schuh is Director, Agriculture and Rural Development, The World Bank, Washington, D.C. Matthew Smith was a research specialist at the University of Minnesota, St. Paul, at the time this paper was written. Maria Ignez Schuh is an independent analyst.

Abstract

Mexico is a nation in the process of change. It is a newly industrialized country with a growing population, now over 70 million. Until getting caught by the 1981 petroleum price decline and sharply higher interest rates on its large international debt, Mexico represented a real economic success story. GNP per capita had increased 3.8 percent per year from 1960 to 1981, even while population was increasing by more than 3 percent annually.

Now, however, Mexico has become one of the major debtor nations of the world. Servicing its debt is a tremendous burden on the economy. Mexico also confronts a huge foreign trade deficit, domestic recession, and potentially disastrous liquidation of many capital assets.

To resolve its debt crisis, and the recession that crisis has created, Mexico clearly must increase foreign exchange earnings. It must participate more effectively in the world economy. After the debt crisis is resolved, continued foreign trade will be critical to sustain domestic economic development.

Revenues from petroleum reserves are helping. The meat industry represents another opportunity to increase foreign exchange. Mexico in the past has been a major exporter of meat products, and has the potential to expand that industry.

Mexico's economic problems, and measures such as devaluation and price controls to deal with those problems, however, are undermining long-term strength and eroding real potential, especially in the agricultural sector.

The devalued peso makes imported feed grain more expensive. Dairy and swine herds have been cut by 40 percent, as has much of the country's poultry flock.

The potential demand for meat exports, however, is substantial. Domestic consumption of meat and meat products in Mexico historically has been high, and will increase with economic strength and a probable shift of income to lower income groups. To meet these increasing demands for meat products, the Mexican meat industry must find a way to increase production at minimal cost.

This chapter details trends of production, consumption, import and export in Mexico's grain and meat industries. It also explores the valuable role that isolated soy protein can play in increasing output of meat products while saving significant quantities of resource inputs.

Introduction

Soy proteins are food ingredients used by food processors when the natural, functional, and organoleptic characteristics of the ingredient are important. They are all based on soy flour - a 50 percent protein product produced by removing the hull and oil from the soybean. Historically, the first products developed were soy flour and textured soy protein. These are produced by respectively, grinding or extruding soy flakes to change their physical form.

More recently, the protein has been extracted or isolated from the flakes to produce isolated soy protein whlch has a minimum of 90 percent protein level on a dry basis. These ingredients are supplied in powdered, granular, and fibrous forms depending upon the functionality desired.

Technology is now available which makes it possible to use isolated soy protein as a nutritional, functional ingredient in meat products. The isolated soy protein is used with meat in meat products, making it possible to increase the consumption of the reformulated meat products 25 percent or more with the same meat supply. The unit cost of the meat products' raw materials may be reduced 10-15 percent or more.

The use of isolated soy proteins would appear to be of considerable value to a middle-income, newly industrialized country like Mexico for two reasons: First, Mexico is in the throes of a major economic crisis as a consequence of its having taken on a large burden of foreign debt (at quite short terms) at the very time that interest rates in international capital markets increased very greatly and the world economy entered the most severe economic slump in the post-World War II period. Austerity measures applied as a condition for receiving assistance from the international community to deal with this crisis created a severe economic recession in Mexico and imposed enormous shocks on the domestic economy. (More detail is provided below.)

To deal with this short-term crisis, Mexico will have to reorient its economy more toward the international economy so as to earn the foreign exchange needed to service its foreign debt and ultimately to repay it. Towards that end, it has already undertaken a massive devaluation of the peso which, if sustained in real terms, will make imports more expensive and exports more competitive.

Mexico has the potential to become a significant exporter of meat products. If it does, increasing the domestic supply of meat products in an efficient manner will have a high payoff to society and reduce the domestic costs of reorienting the economy toward the external market. Similarly, the massively devalued peso makes imports of feed grains more expensive. Hence, it will be more expensive in the future in terms of domestic resource costs for Mexico to import feed grains and produce value added in the form of meat products, either for domestic consumption or export. Anything which lowers the cost of producing this value added will be of considerable value to the domestic economy.

The second reason for which the use of isolated soy protein may be of considerable value to Mexico assumes that it will get through its present debt-induced crisis and once again reestablish previous rates of economic development. If it does that, the demands on foreign exchange may still be severe, even though large deposits of petroleum provide the basis for ample foreign exchange earnings and the peso may recover from its present lows in real terms. The problem is that it is difficult for any domestic agricultural sector to accommodate increases in domestic demand for agricultural products consistent with those that prevailed in Mexico in the past.

Background on Mexico and Mexican Development

Mexico is a country of over 70 million people (71.2 million as of mid-1981) with a GNP per capita in 1981 of $2,250. (By way of contrast, GNP per capita for Brazil in the same year was $2,220, and for the United States it was $12,820.) Mexico has had one of the highest population growth rates in the world - 3.3 percent per year in the 1960's and 3.1 percent in the 1970's. The population growth rate is projected to decline to 2.6 percent per year for the 1980-2000 period, which would give it a population of 91 million by 1990.

Until its recent economic crisis, Mexico was in some respects an econo-mic success story. GNP per capita had grown at a rate of 3.8 percent per year in the period 1960 to 1981 and it did that with modest levels of domestic inflation. When combined with population growth rates of over 3.0 percent per year, this was an unusual economic performance. For that reason, Mexico has been classified as among the newly-industrialized countries (the NIC's).

In the late 1970's, Mexico successfully announced increasingly large international petroleum reserves. In the light of high world prices for petroleum, the discovery of large petroleum deposits held out the promise of a bright future for this country. However, the glitter of that future is now dimmed as Mexico fights its way through a major international financial crisis.

In some respects Mexico's problems started at home. Despite repeated assurances that it would not "Venezuelanize" its economy, the government proceeded to induce growth by expanding demand through deficit spending. This led to an acceleration in the domestic rate of inflation which, with a fixed nominal rate of exchange for the peso, led to an increasingly over-valued currency. This caused imports to burgeon and exports to decline. The result was an increasingly severe balance of payments problem.

To support and maintain its domestic development objectives, the government turned increasingly to the foreign capital markets. The private sector also increased its foreign borrowings since dollar loans were quite attractive at prevailing exchange rates and distortions in the domestic economy.

In 1981, the fall in petroleum prices caused projected revenue to decline. Rather than alter policy by slowing the growth rate or significantly devaluing the peso, the government expanded its borrowing. The peso was permitted to decline very modestly in nominal terms and strong import restrictions were imposed. However, the effects of these restrictions were more than offset by accelerating inflation and a continued high level of government spending.

A massive capital outflow finally induced an equally massive devaluation - from 26.62 pesos in January 1982 to 148.50 at year's end. Foreign exchange controls were imposed, the growth in GNP turned negative, and imports fell from $14.4 billion to $9.5 billion, with 63 percent of the reduction due to cutbacks in the private sector. The downturn in private economic activity caused the deficit to grow to 17.2 percent of the GNP as government spending increased from 38 percent of GNP in 1981 to 51 percent of GNP in 1982.

Clearly, Mexico's economic policies were infelicitous, and it was slow to respond to a rapidly deteriorating situation. However, not all of Mexico's problems were of its own making.[1] Starting in late 1980, the value of the U.S. dollar started on an unprecedented rise which ultimately amounted to 57 percent in real terms by 1983. Although the peso was permitted to decline slowly in nominal terms in the latter half of 1980 and through 1981, it continued to rise in real terms from its already overvalued situation through the end of 1981. The peso became significantly overvalued relative to the U.S. dollar. Moreover, in being essentially pegged to the value of the dollar in nominal terms, the value of the peso rose in real terms relative to the currencies of third countries as the value of the U.S. dollar sped upward. It is little wonder that exports stagnated and import pressures grew. Moreover, even had the peso been floating in this period, the adjustment imposed on the Mexican economy as a consequence of the rise in the value of the dollar would have been very significant.

The great rise in interest rates in the United States and in the international capital market further exacerbated Mexico's problems. Starting in October 1979, the U.S. Federal Reserve stopped monetizing the debt produced in the United States as a consequence of the large Federal deficit. By spring of 1980, nominal interest rates (prime rate) were up above 20 percent. Later, these rates were to rise still higher. Mexico was forced to refinance its already large debt at increasingly higher interest rates, and the cost of servicing the debt grew at a very rapid rate.

By the end of 1982, Mexico had reached an accord with the IMF and its creditors and launched into an austerity program. Government spending is to be reduced in order to bring it down to 8.5 percent of the GNP. Taxes and prices of goods and services supplied by the public sector are to rise. And the growth rate of the economy is expected to decline.

At the time of this writing, the Mexican economy appears to have responded to the orthodox shock treatment and is making a remarkable turnaround.

---

[1]For more detail on this issue, see Schuh, G. Edward, "Monetary Disturbances in a Changed International Economy: The Case of Mexico's Agriculture and Mexican/U.S. Trade," prepared for book on Mexican/U.S. Relations, Stanford University Press.

However, continued problems lurk behind the scene. Domestic price inflation continues high. The peso now appears to be undervalued, in contrast to its chronic overvaluation in the past. This can lead to distortions equally as serious as overvaluation.

In addition, domestic prices for some products have been fixed in an effort to cushion the effects of the devaluations to consumers. Unfortunately, fixing the price in nominal terms led to a dramatic decline in the real value of the prices of these commodities. This has led to a liquidation of some 40 percent of the dairy herd, 40 percent of the swine herd and a large population of the poultry flock. These data suggest that Mexico is re-establishing equilibrium in its external accounts by consuming a significant part of its capital stock. The longer term consequences of this can be quite serious.

Trends in the Meat and Grain Sector

Meats

Data on the production and consumption of meat in Mexico by five-year periods are summarized in Tables 1 and 2, respectively. On a tonnage basis, beef is the most important meat product, although its share of both production and consumption has been declining. Pork has become relatively more important on both the production and consumption side, but poultry has experienced the most rapid growth, with its share of total meat production almost doubling from 1961-65 to 1976-80.

Production of poultry almost tripled over the period while production of pork more than doubled. Beef production experienced a more modest growth, as did sheep and goats.

Mexico has been a significant exporter of meat products over the years (Table 3). Beef has been the major meat export, although exports of this product stagnated in the latter half of the 1970's. Pork became an export item in the 1970's and by the latter half of the 1970's meat exports were significant. At the tail end of the 1970's there were some modest exports of poultry as well, but beef still constituted 95 percent of exports on a tonnage basis in the last half of the 1970's.

Mexico is also a modest importer of meat products (Tables 4 and 5), although overall it is a net exporter of meat products. Poultry has been the largest single meat import over the twenty-year period, but imports of sheep and goat meat have also been significant in both relative and absolute terms. By the latter half of the 1970's Mexico was importing almost two thousand tons of meat a year (in contrast to a total consumption of 1.3 million tons of domestic consumption of meat). Significant quantities of all four meat items were imported in this period.

Data on the consumption of meat products per capita are summarized in Table 6. Beef and veal are the single largest components, but consumption of these items declined over the period on a per capita basis. Per capita consumption of pork increased significantly over the period, as did that of poultry. Consumption of poultry almost doubled.

166

Table 1

Production of Meats, by Category, and Percentage Distribution, 1961-1980, Mexico

Year	Indigenous Pig Meat	Indigenous Beef & Buffalo Meat	Indigenous Mutton & Goat Meat	Poultry	Total Meat
Average Production Per Year (Metric Tons)					
1961-1965	199,940	448,996	24,797	116,841	790,574
1966-1970	225,712	506,500	27,875	194,560	954,646
1971-1975	317,124	527,100	25,592	296,038	1,165,853
1976-1980[1]	426,715	629,860	31,043	381,000	1,468,618
Distribution By Category (Percent)					
1961-1965	25.3	56.8	3.1	14.8	
1966-1970	23.6	53.1	2.9	20.4	
1971-1975	27.2	45.2	2.2	25.4	
1976-1980	29.1	42.9	2.1	25.9	

Sources: Data 1961-1977 from FAO Productions Tapes 1975, 1978 (Paulino-IFPRI).
[1]Preliminary data, 1978-1980 from Production Yearbook, Vol. 34, 1980 and Vol. 35, 1981.

Table 2

Consumption of Meats, by Category, and Percentage Distribution, 1961-1977, Mexico

Average Consumption Per Year (Metric Tons)				
Year	Pig Meat	Veal & Beef	Mutton & Goat	Poultry	Total Meat
1961-1965	200,065	366,982	25,112	117,359	709,518
1966-1970	225,983	404,128	28,644	195,238	853,993
1971-1975	328,358	446,872	27,855	297,146	1,100,231
1976-1977	418,873	505,471	28,901	326,698	1,279,943

Distribution By Category (Percent)				
	Pig Meat	Veal & Beef	Mutton & Goat	Poultry
1961-1965	28.2	51.8	3.5	16.5
1966-1970	26.5	47.3	3.4	22.8
1971-1975	29.8	40.6	2.5	27.0
1976-1977	32.7	39.5	2.3	25.5

Sources: Agriculture Toward 2000 tape, IFPRI.

Table 3

Exports of Meat, Fresh, Chilled and Frozen Mexico, 1961-1965

Average Exports Per Year (Metric Tons)				
Year	Bovine	Sheep & Goat	Pork	Poultry	Total
1961-1965	26,632	0	0	0	26,632
1966-1970	30,857	0	0.8	0	30,858
1971-1975	25,374	0	558	0	25,932
1976-1980	22,607	0	1,251	13	23,871

Average Value of Exports Per Year ($1,000)				
Year	Bovine	Sheep & Goat	Pork	Poultry	Total
1961-1965	17,847	0	0	0	17,847
1966-1970	32,276	0	8	0	32,284
1971-1975	36,329	0	633	0	36,962
1976-1980	36,278	0	2,043	84	38,405

Source: Data for all trade 1961-1963 FAO Trade Yearbook, Vol. 20, 1966, Rome.
1964-1969 FAO Yearbook Vol. 24, 1970.   1977-1978 FAO Yearbook Vol. 33, 1979.
1971-1973 FAO Yearbook Vol. 28, 1974.   1979-1980 FAO Yearbook Vol. 35, 1981.
1974-1976 FAO Yearbook Vol. 30, 1976.   *Data for swine for 1980 are unofficial figures.

Table 4

Imports of Meat, Fresh, Chilled and Frozen Mexico, 1961-1965

Average Exports Per Year (Metric Tons)					
Year	Bovine	Sheep & Goat	Pork	Poultry	Total
1961-65	22	0	12	37	71
1966-70	107	167	44	65	383
1970-75	79	533	44	852	1,508
1976-80	327	497	280*	717	1,821

Average Value of Exports Per Year ($1,000)					
Year	Bovine	Sheep & Goat	Pork	Poultry	Total
1961-65	28	0	9	32	69
1966-70	122	98	19	48	287
1971-75	61	366	25	480	932
1976-80	487	722	260*	448	1,917

Source: see Table 3.

Table 5

Distribution of Meat Imports by Category, Mexico, 1961-80 (Percent)

Year	Bovine	Sheep & Goat	Pork	Poultry	Total
1961-1965	31	0	17	52	100
1966-1970	28	44	11	17	100
1971-1975	5	35	3	57	100
1976-1980	18	27	16	39	100

Source: Table 4.

Table 6

Consumption of Meat Per Capita, Average Per Year, Mexico, 1961-1977 (kilograms)

Year	Pig Meat	Beef & Veal	Mutton & Goat	Poultry	Total
1961-1965	4.98	9.13	0.62	2.92	17.65
1966-1970	4.78	8.55	0.61	4.13	18.07
1971-1975	5.91	8.05	0.50	5.35	19.81
1976-1977	6.78	8.12	0.46	5.25	20.61

Source: Calculated from Table 2 and Population Data.

Livestock Products

Data on the production and consumption of livestock products are summarized in Tables 7 and 8. Production of milk and dairy products almost tripled during the twenty-year period, and production of eggs approximately tripled. The bulk of the dairy products is consumed in the form of fresh milk, and consumption increased at approximately the same rate as production.

Table 7

Production of Livestock Products, Average Per Year,

Mexico, 1961-1980

Year	Cow Milk (Whole & Fresh)	Evap. Condens. Milk	Dry Whole Cow Milk	Skim Milk, Buttermilk, Dry
	1,000 m.t.	m.t.	m.t.	m.t.
1961-1965	2,305	40,719	5,634	1,965
1966-1970	2,716	72,400	11,100	—
1971-1975	3,438	95,845	14,820	1,320[1]
1976-1980	6,178	151,155	25,681	4,848

Year	Cheese (all kinds)	Butter & Sheep	Hen Eggs
	m.t.	m.t.	m.t.
1961-1965	61,948	12,600	169,845
1966-1970	—	18,000[1]	217,200
1971-1975	80,755[1]	21,000[1]	399,524
1976-1980	94,185[1]	24,560[1]	507,932

Source: FAO Production Yearbook, Vols. 24-35, 1970-1981.

[1] FAO estimates.

Table 8

Total Consumption of Livestock Products, Average Per Year,

Mexico, 1961-1980

Year	Milk, Fresh	Evap. Cond. Milk	Dry Milk
	m.t.	m.t.	m.t.
1961-1965	2,305,000	43,919	26,482
1966-1970	2,716,000	84,245	36,427
1971-1975	3,439,139	110,736	66,460
1976-1980	6,178,806	172,616	114,119

Year	Butter	Cheese (all kinds)	Hen Eggs
	m.t.	m.t.	m.t.
1961-1965	12,944	62,128	169,846
1966-1970			284,374
1971-1975	25,363	81,625	399,254
1976-1980	41,839	95,976	508,837

Source: Production and Imports — Exports.

Table 9

Imports of Livestock Products, Average Per Year,

Mexico, 1961-1980

Year	Fresh Milk	Dry Milk	Condensed & Evap. Milk	Butter	Cheese & Curd	Eggs in Shell
Metric Tons						
1961 - 1965		20,848	8,000[1]	430[2]	180	20
1966 - 1970		25,335	11,849	1,004	572	15[3]
1971 - 1975	739	51,820	14,930	4,363	872	27
1976 - 1980	1,006	88,438	21,292	16,679	1,817	1,270
Value ($1,000)						
1961 - 1965	—	5,039	2,116[1]	316[2]	179	19
1966 - 1970	—	7,533	4,081	802	789	25[3]
1971 - 1975	244	37,235	6,663	5,730	1,223	93
1976 - 1980	486	56,562	13,072	24,847	3,780	1,713

Source: FAO Yearbook, Vols. 20, 24, 26, 28, 30, 33, 35, (1966, 1970, 1972, 1974, 1976, 1979, 1981).

[1] Data available only for 1964 and 1965.

[2] Data not available for 1962.

[3] Data not available for 1966.

Data on the imports of livestock products are summarized in Table 9. although on a small base, imports of livestock products increased very significantly over the period. Imports of dairy products have benefitted from export subsidies of exporting countries. In the latter half of the 1970's, total imports of livestock products were averaging approximately $100 million.

Prices of Meat Products

Wholesale and retail prices for the four meat components are summarized in Tables 10-13. Relative prices shifted rather significantly among the various components over the period. As an average for the five-year periods, the U.S. wholesale price index increased 111 percent from 1961-65 to 1975-80. Beef prices in Mexico (in dollar terms) increased 160 percent in that same period, thus experiencing a significant increase. Pork prices, on the other hand, increased 118 percent, which was only a slight rise in real terms.

Table 10

Beef Prices, Mexico, 1961 - 1980 (Per Kilo)

Year	Producer (wholesale, carcass)[a]		Retail[b]	
	Pesos	Dollars[1]	Pesos	Dollars[1]
1961 - 1965	7.24	0.58	13.57	1.08
1966 - 1970	10.18	0.81	16.53	1.32
1971 - 1975	16.00	1.28	27.45	2.20
1976 - 1980	34.80[2]/	1.56	65.77	2.95

Sources: [a]1960 - 1971, Revista de Estadistica (T-DAM file:); 1972 - 80, SARH, DGEA Estat. Subsection Recuario, 1972 - 1977 and 1980.

[b]1960 - 1973, Revista de Estadistica (T-DAM file); 1974 - 1980 Donnat records (Vacla notebook).

[1]Used nominal exchange rate IMF. International Financial Statistics.

[2]Excludes data of years 1978 and 1979.

Table 11

Pork Prices, Mexico, 1961 - 1980 (Per Kilo)

Year	Producer (wholesale, carcass) [a]		Retail [b]	
	Pesos/Kilo	Dollars/Kilo[1]	Pesos/Kilo	Dollars/KILO[1]
1961 - 1965	10.10	0.81	16.27	1.30
1966 - 1970	12.91	1.03	18.45	1.48
1971 - 1975	22.12	1.77	29.30	2.34
1976 - 1980	-----		66.07	2.96

Sources: [a]1961 - 1965, Revista de Estadistica (T-DAM file:).

[b]1961 - 1973, Revista de Estadistica (T-DAM file), 1974 - 1980 Donnat records (Vacla notebook).

[1]Nominal exchange rate IMF.

175

Table 12

Poultry Prices, Mexico, 1961 - 1980 (Per Kilo)

Year	Producer (live weight)[a]		Wholesale (slaughter price)[b]	
	Pesos	Dollars	Pesos	Dollars
1961 - 1965	----	----	----	----
1966 - 1970	----	----	----	----
1971 - 1975	11.67[1]	0.93	20.43[2]	1.63
1976 - 1980	23.64	1.06	35.45[3]	1.59

Sources: [a]Union Nactional de Aviculturs, published in *Boletin Interno* No. 51, Dec. 17, 1980 (1973 - 1978) SARH, DGEA and DAVEP.

[b]SARA, DGEA, Estat. del Subsector Pecuario de los EE. UU Mexicanos.

[1]Includes years 1973 - 1975.

[2]Includes years 1972 - 1975.

[3]Includes years 1976, 1977, 1980.

Table 13

Mutton Prices, Mexico, 1961 - 1980 (Per Kilo)

Year	Wholesale (carcass weight) [a]		Retail (carcass weight) [b]	
	Pesos	Dollars	Pesos	Dollars
1961 - 1965	10.79	0.86	15.91	1.27
1966 - 1970	13.25	1.06	19.59	1.57
1971 - 1975	23.51	1.05	31.27	1.40
1976 - 1980	----	----	----	----

[a]1961 - 1965, Revista de Estadistica (T-DAM file).

[b]1961 - 1975, Revista de Estadistica (T-DAM file).

The wholesale price index increased 46 percent from 1971-75 to 1976-80. Poultry slaughter prices actually declined 2.5 percent in that same period, hence in real terms they experienced almost a 50 percent decline. The U.S. wholesale price index increased 42 percent from 1961-65 to 1971-75. Mutton prices increased only 22 percent, hence they, too, experienced a decline in real terms.

Overall, poultry meat is displacing other forms of meat due to the adaption of modern technology and a large increase in output. It is these developments that are causing a decline in the relative price of this product.

The Grain Sector

Data on the production of grain in Mexico are presented in Table 14. On a value basis, maize is by far the most important grain produced in Mexico, although it declined significantly in relative importance in the twenty-year period being considered. Maize is used for both human consumption and for feeding to livestock, when it is used directly on the farm. However, it is not allowed as an ingredient in commercial feeds.

Table 14

Average Annual Grain Production

Mexico, 1961-1980

	Wheat	Rice Milled	Maize	Sorghum & Millet	Others	Total
Thousand Metric Tons						
1961-1965	1,672	220	7,369	452	254	9,956
1966-1970	2,171	271	8,845	2,083	280	13,650
1971-1975	2,264	340	8,783	3,141	388	14,916
1976-1980	2,718	335	9,658	4,228	544	17,483
Percent of Grain Production						
1961-1965	16.8	2.2	74.0	4.5	2.5	100
1966-1970	15.9	2.0	64.8	15.3	2.0	100
1971-1975	15.2	2.3	58.8	21.1	2.6	100
1976-1980	15.5	1.9	55.2	24.2	3.2	100

Source for Grain Production: FAO, "Production Yearbook Tapes, 1975, 1979 and 1980." (IFPRE); FAO, "Global Agricultural Programming System Supply Utilization Accounts Tape." June, 1980.

Total production of grain increased almost 70 percent from the first five-year period to the last. Sorghum was by far the most dynamic sector, increasing almost ten-fold, and was accounting for approximately 25 percent of total grain production by the end of the 1970's.

Data on the utilization of grains as feed are summarized in Table 15. As these data indicate, there has been a major transformation in the utilization of grain for feed. In the early 1960's, maize was the major feed grain. By the end of the 1970's, however, it had been displaced from that position by sorghum. Wheat declined in relative importance as a feed grain starting in the latter half of the 1960's, but stayed at a relatively constant proportion thereafter.

Table 15

Grain Utilization as Feed

Mexico, 1961-1980

	Wheat	Maize	Sorghum & Millet	Others	Total
Thousand Metric Tons					
1961-1965	387	770	495	78	1,731
1966-1970	518	1,085	1,913	65	3,577
1971-1975	705	1,318	3,111	146	5,280
1976-1980	853[1]	1,255[1]	3,538[1]	175[1]	5,821

Percent of Grain Utilization as Feed

	Wheat	Maize	Sorghum	Others
1961-1965	22.4	44.5	28.6	4.5
1966-1970	14.4	30.3	53.5	1.8
1971-1975	13.3	25.0	58.9	2.8
1976-1980	14.6	21.6	60.8	3.0

Source for Grain Utilization: (FAO Production Yearbook Tapes, 1975, 1979 and 1980); FAO, "Global Agricultural Programming System Supply Utilization Accounts Tape." June, 1980.
[1] Preliminary for 1980.

Total utilization of grain in livestock feeding increased over threefold during the period considered. This is associated with the rapid expansion of the swine and poultry sectors, and especially to the expansion of the latter. Only a small proportion of slaughtered beef is fed. Most beef is raised on grass.

Mexico's imports of grain have increased very rapidly during the 1970's (Tables 16 and 17). This growth in imports has been primarily in the form of feed grains for the rapidly growing swine and poultry sectors. Imports of soybeans and other protein sources also increased significantly during the period, with these components used in livestock and poultry rations. During the latter half of the 1970's, imports of grains were averaging over a half billion dollars a year.

Table 16

Imports[1] of Grain, Average Per Five-Year Period

Mexico, 1961-1980 (1,000 metric tons)

	Wheat & Flour, Wheat Equiv.	Rice	Barley, Rye, & Oats	Maize	Other Cereals	Total
1961-1965	28	4	41	128	57	257
1966-1970	—	6	14	157	25	201
1971-1975	520	22	76	1,053	310	1,980
1976-1980	591	26	62	1,706	1,058	3,443

Source: FAO Trade Yearbook, Vol. 26, 1972; Vol. 28, 1974; Vol. 30, 1976; Vol. 35, 1979; Vol. 35, 1981.

[1] Dates for imports include FIL international free zone.

Table 17

Value and Percentage Distribution of Imports,

Mexico, 1961-1980

	Wheat & Flour, Wheat Equiv.	Rice	Barley, Rye, & Oats	Maize	Other Cereals	Total
Value ($1,000)						
1961-1965	2,850	482	3,425	9,356	3,352	19,465
1966-1970	80	893	1,154	20,003[1]	2,176	24,306
1971-1975	68,551	7,665	16,006	148,235	45,785	286,242
1976-1980	96,564	9,695	11,454	250,630	139,065	507,408
Percentage Distribution						
1961-1965	10.7	1.6	15.8	49.6	22.3	100
1966-1970	0.1	2.8	6.8	77.9	12.4	100
1971-1975	26.2	1.1	3.8	53.2	15.7	100
1976-1980	17.2	0.7	1.8	49.6	30.7	100

[1] Does not include data for 1967 and 1968.

At one time Mexico was a significant exporter of grain (Tables 18 and 19). In fact, in the latter half of the 1960's it was exporting on average over a million tons a year (to be a net exporter) and earning significant amounts of foreign exchange earnings. The major cereal export at that time was maize.

Table 18

Imports[1] of Grain, Average Per Five-Year Period

Mexico, 1961-1980 (metric tons)

	Wheat & Flour, Wheat Equiv.	Rice	Barley, Rye, & Oats	Maize	Other Cereals	Total
1961-1965	129,903	16,538[1]	656[a]	326,542	395	474,034
1966-1970	261,712	45,729[2]	—	758,063	141,535	1,207,039
1971-1975	32,371	9,312[3]	5,023[5b]	146,003	14,804[8]	207,613
1976-1980	17,920	16,595[4]	12,582[6b]	642[7]	1,165	48,904

[1]Does not include data of 1964.
[2]Only data of 1968 is included.
[3]Does not include data of 1971, 1975.
[4]Does not include data of 1980.
[5]Does not include data of 1971.

[6]Does not include data of 1976.
[7]Does not include data of 1978.
[8]Does not include data of 1974.
[a]Includes only data on oats.
[b]Includes only data on barley.

By the end of the 1970's, exports had declined very significantly and the country had become a net importer of grains on a significant scale. The largest decline was in maize, and associated in part with the rapid expansion in the use of feed grains. However, it should be noted that the peso became significantly overvalued in the latter half of the 1970's as domestic price

Table 19

**Value of Grain Exports and Percentage Distribution,
Mexico, 1961-1980**

	Wheat & Flour, Wheat Equiv.	Rice	Barley, Rye, & Oats	Maize	Other Cereals
Value ($1,000)					
1961-1965	16,516	2,288[1]	105[a]	23,344	77
1966-1970	10,308	3,743[2]	—	41,899	7,111
1971-1975	19,413	695[3]	289[5b]	8,160	842[7]
1976-1980	4,292	4,395[4b]	1,390[6b]	460	559
Percentage Distribution					
1961-1965	27.4	3.5	0.1	68.9	0.1
1966-1970	21.7	3.8	—	62.8	11.7
1971-1975	15.6	4.5	2.5	70.3	7.1
1976-1980	36.6	33.9	25.8	1.3	2.4

[1] Does not include data of 1964.
[2] Included only data of 1968.
[3] Does not include data of 1971, 1975.
[4] Does not include data of 1980.
[5] Does not include data of 1971.
[6] Does not include data of 1976.
[7] Does not include data of 1978.
[8] Does not include data of 1974.
[a] Includes data only for oats.
[b] Includes data only for barley.

inflation burgeoned out of control and Mexico tried to retain a fixed exchange rate regime. The effects of a significant devaluation of the peso in 1976 were rapidly eroded due to the failure to pursue appropriate complementary monetary and fiscal policies. An overvalued currency is an implicit import subsidy and an implicit export tax. Hence, the distorted value of the currency not only provided strong incentives for imports to grow, it also reduced the incentives to product maize and other cereals domestically.

Total Trade in Agricultural Products

Overall, Mexico still earns a modest surplus on its agricultural trade account (Tables 20 and 21). This surplus declined very significantly from the latter half of the 1960's to the latter half of the 1970's, however. Moreover, there has been a very significant shift in the composition of the balance. Almost all components of imports increased, and increased very rapidly, with imports of feed grains for the rapidly growing swine and poultry sectors leading the way. Exports of cereals declined very significantly, however, while fresh fruits and vegetables increased very significantly. Exports of live animals and meat products increased modestly.

Table 20

Total Trade in Agriculture Products,

Mexico, 1961 - 1980 ($1,000)

Years	Agri-Products Total	Food & Animals	Meat & Meat Prep.	Dairy Prod. of Eggs	Cereals & Prepared	Oil Seeds
Value of Imports						
1962 - 1965	106,325	53,550	775	6,150	22,550	700
1966 - 1970	126,269	54,492	3,488	10,336	17,143	5,515
1971 - 1975	606,496	433,828	7,561	51,122	284,069	37,641
1976 - 1980	1,330,720	861,559	23,457	100,497	508,570	218,621
Value of Exports						
1962 - 1965	536,150	354,350	22,250	---	46,525	4,675
1966 - 1970	587,522	414,756	34,991	161	58,188	4,554
1971 - 1975	917, 267	681,448	40,953	368	17,535	9,624
1976 - 1980	1,631,904	1,172,821	44,262	394	14,087	40,249

Source: FAO Tradebook, Vols. 22, 24, 26, 30, 32, 35, 1968 - 1981.

Table 21

Agriculture Trade Balance, Mexico, 1961 - 1980 ($1,000)

Years	Agri-Products Total	Food & Animals	Meat & Meat Prep.	Dairy Prod. of Eggs	Cereals & Prepared	Oil Seeds
1962 - 1965	429,825	300,800	21,475	-6,150	23,975	3,975
1966 - 1970	461,253	360,264	31,503	-10,175	41,045	-961
1971 - 1975	310,771	247,620	33,392	-50,754	-266,534	-28,017
1976 - 1980	301,184	311,262	20,805	-100,103	-494,483	-178,372

Source: Exports - Imports, Table 19.

Synthesis, Summary and Conclusions

Mexico is now at a crossroads. Its future will be dramatically different than its past. If it is to service its large foreign debt while at the same time amortizing it, it will need to import significantly less in the future in a relative sense and export more. An important mechanism for bringing this shift about will undoubtedly be a realignment in the value of the Mexican peso. Granted that the peso is undoubtedly undervalued at the present time, the real value of the peso is still likely to be significantly less (in terms of the dollar and other currencies) in the remainder of this decade than it was in the latter half of the 1970's. The only thing that could significantly alter that judgment is if there should be a major conflagration in the Middle East which would send petroleum prices shooting upward again.

If the real value of the peso is in fact lower through the remainder of this decade, as expected, it will have significant effects on agriculture and on consumers. In the first place, the cost of grain imports will rise significantly. Domestic grain prices should rise, and eventually this will lead to an increase in domestic production. In a relative sense, prices of grain will rise, however.

At the same time, Mexico's exports of meat should be more competitive in the international market. This will be reflected in higher relative prices in the domestic economy as the economy exports its domestic supplies. The

185

production of livestock and meat products will undoubtedly remain profitable, since the price of meat products and grains should rise in approximately the same proportions. However, as tradeables, the price of both should rise relative to non-tradeable goods. This will put a burden on consumers at all income levels. Consequently, the adaption of a new technology which extends the supply of meat while at the same time reducing its cost should be of considerable value to the economy, whether it does it by increasing exports and thereby earning more foreign exchange or by increasing supplies for domestic consumption, thereby helping to keep food costs lower than they would otherwise be, or both.

Projected Impact of Isolated Soy Protein Food Ingredients on Meat Resources Used for Production of Processed Meats

While a complete financial or micro-economic evaluation of isolated soy protein and the processed meat industry is a subject for another paper, a review here of the technical coefficients is quite helpful. This review will provide an indication of the order of magnitude of the opportunity provided by isolated soy protein.

For example, a typical emulsified meat product contains 12% protein, 24% fat, 61% moisture and 3% salt, spices and cures. The meat used for such a product is typically 70% lean and 30% fat. Using traditional technology, 100 kg of meat produces 125 kg of processed meat products such as mortadella. Alternatively, 100 kg of meat in combination with 4 kg of isolated soy protein and 7 kg of fat will produce 154 kg of processed meat products — an increase of 23 percent. Isolated soy protein, when properly hydrated, provides the same combination of protein and moisture as lean meat. The final consumer product maintains the same traditional quality, both nutritionally and sensorally.

For every 100,000 tons of emulsified meat products produced in Mexico, the potential implication of using isolated soy protein are of the following magnitude:

- 100,000 tons of emulsified meat requires:

- 80,000 tons of meat using traditional technology

- This same 80,000 tons of meat with 3,200 tons of isolated soy protein and 5,600 tons of fat would produce 123,200 tons of emulsified meat products.

- To produce the same 123,200 tons of emulsified meat products with traditional technology would require 98,600 tons of meat — or an additional 18,600 tons of meat (98,600-80,000) versus an additional 3,200 tons of isolated soy protein and 5,600 tons of fat.

- It would require 3 times this additional amount to provide the Mexican population with an additional 1 kg of meat products per capita per year (123,200-100,000) X 3 = 69,600 tons.

186

- To provide the additional kilo per capita per year would require 55,800 additional tons of meat (18,600 x 3) or on a carcass basis 79,900 additional tons. Table 1 shows 1976-80 Mexican production, on a carcass basis, to be 629,860 tons of beef and 426,715 tons of pig meat. The increment is significant.

- Such an additional quantity of meat would require slaughtering, per year, an additional:

    285,000 head of cattle (400 kg each)
    or
    1,427,000 head of hogs (80 kg each)

- This same increment to provide the additional kilo per capita per year, through the processed meat sector, could be achieved with 9,600 tons of isolated soy protein (3,200 x 3).

The magnitude of the opportunity suggested by these technical coefficients indicates that it is well worth pursuing.

SOME ADDITIONAL ISSUES

A number of other issues augment the potential value of isolated soy protein to the Mexican economy. First, Mexico has a serious infrastructure problem. This includes not only meat/meat processing, but also the movement and storage of grains.

Consider the movement of grains only. One railroad car will hold 40 tons of isolated soy protein. Each ton of soy protein is the equivalent of 4 tons of lean meat or 5 tons of meat with fat, assuming fat is in excess. That is equivalent to 10 tons of hogs live weight, which is equivalent to 45 tons of commercial animal feed assuming a feed conversion of 4.5:1.[2] If milo is 74.9 percent of the formulation and soybean meal 16.5 percent, it will take 12 cars of milo plus 4 cars of soybean meal to produce the same amount of meat. Thus the railroad car ratio is 16:1.

Second, the retrenchment Mexico has gone through has undoubtedly left the poor and middle-income groups now worse off in real terms than they were a few years ago. Reductions in food prices by the use of soy protein would improve the incomes and welfare of these groups. Moreover, the use of soy protein to augment regular meat supplies holds out that promise that overall meat supplies can be maintained or increased in the face of severe economic problems.

Finally, the reduction in livestock herds as a consequence of current economic difficulties will make it difficult to meet combined domestic and export markets in the near future. The use of soy protein would make this goal more feasible, and without a severe resource drain. Savings could occur in both the amount of foreign exchange and investment required.

---

[2]This is a high conversion and assumes feeding the whole herd, not the conversion for just feeding feeder pigs to market weight.

Chapter 6

Quantifying the Profitability of New Technology

*F. H. Schwarz**

I would like to acknowledge, with deep appreciation, the excellent guidance and helpful critiques of Professors Walter P. Falcon and Terry Sicular in both the formulation and drafting of this manuscript and the help of Axel Schwarz in its production.

*F. H. Schwarz is President, International Business Development Corporation, St. Louis, Missouri.

189

Abstract

Financial and economic benefits ultimately become the bases for adoption or rejection of new technology. This chapter details potential financial and economic benefits from utilization of isolated soy protein ingredients in manufacturing processed meat products. Through exploration of two alternatives for increasing meat production, the author demonstrates a methodology for quantifying these benefits.

The first alternative uses traditional meat production technology applied through a program of increasing the size of a beef cattle herd, and, therefore, meat output, by 20 percent over a five-year period.

The second approach does not increase the size of the herd, but incorporates isolated soy protein food ingredients in processed meat products to achieve the same 20 percent increase in total meat output. Incorporating isolated soy protein effects a 26.1 percent increase in processed meat production, for a 20 percent gain in total meat output.

The methodology described here is valuable for both private industry and government policy decision making. Financial profitability suggests that private industry should adopt the technology. Economic (social) profitability warrants government support.

Appropriate government policies to support the use of isolated soy protein may include food composition and labeling regulations, trade regulations and nutritional programs, including government feeding programs.

This thorough analysis considers all meat production factors for both traditional and alternative technologies. It includes the effects of price distortions, wage rates, raw material costs, revenues, cash flows, and additional investments required at all levels — cattle raising, slaughtering and meat processing.

In the example given, the use of isolated soy protein as an ingredient in processed meat products is demonstrated to be intrinsically profitable, both financially and economically. It is also more profitable than increasing cattle output by an equivalent amount.

Net present values of the alternative technologies are calculated for comparison.

The new production goal of increasing total meat (including processed meat products) output by 20 percent is not recognized under the traditional technology until the fifth year. Under the alternative of using isolated soy protein ingredients in processed meat, the increase is achieved in the first year of the program.

Incorporating isolated soy protein in processed meat products demonstrates clear and significant financial and economic advantages in the example given. This alternate technology signficantly raised profits for both private industry and the economy as a whole.

191

Introduction

Background

Within this volume, the paper by Sarma has shown the need for increased supplies of meat and meat products. Smith, Sicular, and Schuh, in the three case studies, demonstrated the value of considering a food ingredient — isolated soy protein — as a means of increasing the supplies of meat products. Sellers and Wolf have shown this ingredient to be well accepted by consumers and technically superior to other protein food ingredients.

What remains to be seen is whether the use of isolated soy protein and the technology it represents are financially profitable for the meat industry and economically profitable for the country in which it is being used. These are empirical questions, the answers to which can be quantified country by country. In addition, the profitability of using isolated soy protein can be compared to the profitability of other means of increasing the meat supply. The purpose of this chapter is to describe the methodology to quantify the profitability of the various alternative means of increasing the meat supply. The methodology commonly employed in project analysis is useful for this application as described here.

Comparing Technologies

Two alternatives will be shown as examples to describe the methodology. The first alternative, designated "current technology," is to increase the size of a given cattle herd, in this case by 20 percent, and therefore, the output of that herd. The outputs of the slaughtering operation and meat processing operation are correspondingly increased 20 percent. Meat processing is defined here as the production of sausages such as bologna or mortadella. A schematic representation of these operations is shown in Figure 1. For illustrative purposes an original herd of 100 brood cows is assumed.

The second alternative, designated "alternative technology," is to maintain the same size cattle herd and slaughtering operation. Use of isolated soy protein in combination with the existing meat supply can increase the output of processed meat products by 26.1 percent. This output, in combination with retail cuts from the slaughterhouse, has the effect of increasing the total meat supply 20 percent. The two alternatives thus accomplish the same objective as shown in Figure 1.

The Meaning of Profitability

The methodology described here is commonly used for analyzing investment alternatives. In the application described in this paper, alternative means of expanding an existing "project" are being compared. Profitability is, therefore, defined as the discounted cash flow of each alternative considered. Any alternative whose net present value is positive is viable. The alternative with the highest net present value is preferred.

Figure 1

Alternatives to Increase the Supply
of Meat and Meat Products

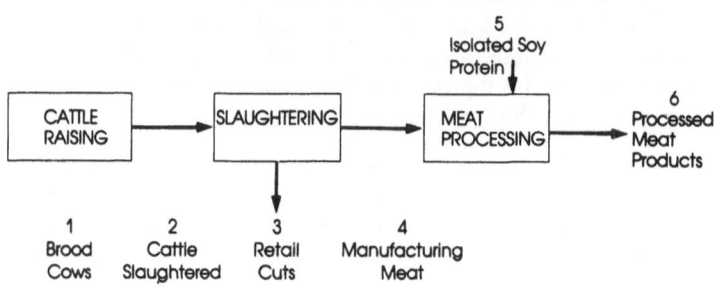

	Original Level	Current Technology Increase Herd Size Twenty Percent	Alternative Technology Use Isolated Soy Protein
1 Brood Cows	100	120	100
2 Cattle Slaughtered (head per year)	79	95	79
3 Retail Cuts (tons per year)	10.7	12.9	10.7
Percent Increase	--	20%	-0-
4 Manufacturing Meats and Fat (tons per year)	13.5	16.2	13.5
Percent Increase	--	20%	-0-
5. Isolated Soy Protein (tons per year)	-0-	-0-	0.5
6. Processed Meats (tons per Year)	34.4	41.3	43.5
Percent Increase	--	20%	26.1%
3 + 6 Retail Cuts Plus Processed Meats (tons per year)	45.1	54.2	54.2
Percent Increase	--	20%	20%

194

An evaluation of private profitability will determine the value of the alternatives to the meat industry. All receipts and payments are valued at market prices. Subsidies are considered as receipts and taxes as payments. No interest or principal payments will be considered, since profitability is determined independently of the method of financing.

An economic evaluation will determine the value of the alternatives to the country. For this evaluation, all price distortions must be eliminated. Distorted prices have the effect of subsidizing or taxing one group in the economy while producing the opposite effect on another group. For example, buyers are subsidized by controlled prices while sellers are taxed. Protected domestic prices, controlled prices, minimum wage rates, and overvalued exchange rates are examples of distorted prices. This analysis is of particular interest to government officials since it shows the profitability of business alternatives to the country as a whole. Economic evaluations are also known as social cost-benefit analysis. These evaluations result in social profitability and are based on shadow prices, which will be described later.

Policy Implications

If an alternative is both financially and socially profitable, it is well worth pursuing from both a private and governmental standpoint. If it is only financially profitable, it may be pursued privately, but its viability depends on, the policies and price distortions that support it. If it is socially profitable, it is deserving of government support.

Such support for the use of isolated soy protein can be achieved through policies adopted or promoted by the government. At the most basic level, food composition and food labeling should permit the use of this ingredient to the extent that it does not alter the traditional sensory and nutritional qualities of the food products in which it is incorporated. The maximum percent at which soy protein isolate can be incorporated in a given product will be used as a technical coefficient in this analysis.

Trade regulations and nutritional programs can be important policy instruments. The ability to secure import licenses readily, the application of preferential exchange rates, and the elimination of customs tariffs and import taxes are primary examples. Employing this ingredient in key governmental programs such as maternal-child health programs, school lunch programs, and other institutional feeding programs could be beneficial. The value of imported isolated soy protein can be judged by competition in the marketplace if governments permit market access to this product without imposing price distortions.

One limitation of this methodology is its inability to rank alternatives according to their ability to achieve various specific economic objectives. Examples of these objectives include economic growth, increasing employment, improving foreign exchange reserves, and lowering food prices. The methodology used here can quantify the effect of each alternative towards meeting each objective. It cannot, however, determine whether the combination of these individual effects for one alternative is better or worse than the combination of these individual effects for another alternative. To answer such a distributional question, country officials would need to quantify the relative importance of each of the specific objectives. Establishing the relative importance is generally not very feasible. Because of the difficulties in quantifying distributional differences and in accordance with the generally accepted methodology for project analysis, this chapter will deal only with the profitability of each alternative.

The analysis that follows will first describe the technical coefficients, then the production costs and finally the analysis of cash flow and profitability.

It must be realized that what follows is only an illustration of how to carry out the analysis. It is not meant to imply that what is shown here as most appropriate is necessarily true in all cases. The ranking of the profitability of the alternative depends upon the technical coefficients, prices and government interventions. Each of these areas varies considerably from country to country and from case to case within countries. Hence the need for good analysis.

Data used in this example is representative of a case in Latin America. The analysis in the appendix demonstrates the sensitivity of profitability to changes in selected technical coefficients and prices in this example.

Business experience has demonstrated that the use of isolated soy protein in meat products is financially profitable in 37 countries with meat processing industries. Corresponding economic evaluations have not been made. The reality of the world market, however, speaks well for the value of this ingredient.

Technical Coefficients

The bases for the full analysis are the technical coefficients that describe the relationships between the quantity of various inputs and the quantities of output in a given time period. Different technologies are described by different input: output relationships. Thus, a traditional technology will have a different set of technical coefficients from a more modern technology. The basic technical coefficients for this example are shown in Tables 1, 2, and 3.

Cattle Raising

Table 1 provides examples of coefficients for cattle raising based upon a herd of 100 brood cows. A calf crop per year equal to 80 percent of the brood cows, a feed conversion of 10:1 and an age at market of three years indicates that this is a more traditional technology. A more modern technology would provide a calf crop of 90 percent, a feed conversion of 8:1 and an age of market at two years.

This herd of 100 brood cows supports a total herd of 343 animals. Assuming that cull bulls are not marketed, 79 head are slaughtered per year yielding a total live weight of 35.6 metric tons. A more detailed analysis would include an annual mortality factor not shown here.

Table 1
Technical coefficients
Cattle Raising

Per 100 Cows

Number of cows			100	
Number of bulls			3	
Calf crop			80%	
Cow replacement per year			20%	
Years in breeding herd			4 years	

	Male		Female
Calves per year	40		40
Replacement per year	1		20
Available per year as market animals	39		20

Breeding stock		
Cows	100	
Bulls	3	
Calves 0-1 year	21	
1-2 years	21	
2-3 years	21	
Stock for market		
0-1 years	59	
1-2 years	59	
2-3 years	59	
Total herd	343	
Market animals sold per year	59	
Cull cows sold per year	20	
Bulls retired per year	1	
Total calves per year	80	
Age at market	3 years	
Weight at market	450 kg	
Weight gain on feed	400 kg	
Feed conversion	10:1	
Feed consumption per head 0-3 years	4,000 kg	
Average tons of feed per year	4 ÷ 3 = 1.33 tons/head	

Tons (live weight) sold per year:

Market animals	(59 head)	26.6
Cull cows	(20 head)	9.0
Total	(79 head)	35.6

Labor inputs are shown in Table 4.

Slaughtering

Moving to the next step in Table 2, a 450 kg animal, live weight, yields a carcass of 307 kg, and hide, shank, blood, and offal of 143 kg. The carcass, in turn, is divided as 136 kg of retail meat cuts plus 136 kg of manufacturing meat and 35 kg of fat which will be used as ingredients in processed meats.

Table 2
Technical Coefficients
Slaughtering, Breaking, and Boning

Per Head		
Live weight		450 kg
Carcass weight		307 kg
Retail cuts - with bone	136 kg	
Manufacturing meat (80% lean) - boneless	136 kg	
Fat	35 kg	
Hide, shank, blood, and offal		143 kg

Labor inputs are shown in Table 5.

In establishing technical coefficients it is important to account for the balance between inputs and outputs. If a balance is not achieved at this stage, important costs and revenues may be inadvertently overlooked at a later stage.

Meat Processing

The 136 kg of manufacturing meat and 35 kg of fat are combined with 265 kg of other ingredients to produce 436 kg of processed meats as shown in Table 3. When these are added to the 136 kg of retail cuts, the total supply of meat products is 572 kg per head of cattle slaughtered.

Alternatively, the 136 kg of manufacturing meat and 35 kg of fat plus 15.7 kg of fat purchased in the market (35.0 + 15.7 = 50.7 kg fat total) may be combined with 6.9 kg of isolated soy protein, 20.5 kg of additional water and 335.9 kg of other ingredients to produce 550 kg of processed meats. This provides a meat supply of 686 kg per head of cattle slaughtered or a 20 percent increase because of alternative meat processing technology with the same size cattle herd.

This alternative technology will be compared to using current technology and simply increasing the herd size 20 percent. A detail in the production costs will precede the description of the technical coefficients for an increase in the herd size.

Table 3
Technical Coefficients
Meat Processing

Per Head	Current Technology (all meat)	Alternative Technology (with isolated soy protein
Manufacturing meat (80% lean)	136 kg	136 kg
Fat	35	50.7
Isolated soy protein	--	6.9
Water	--	20.5
Other	265	335.9
Total	436 kg	550 kg
Retail cuts	136	136
Total meat/meat products	572 kg	686 kg

Labor inputs are shown in Table 6.

Production Costs

Production costs and revenues can be identified at the same time the analyst is investigating the technical coefficients. Such costs should come from accounting records but may need to be estimated if good records are not available. As with technical coefficients, all costs, revenues, and profits should be identified and balanced.

Inasmuch as this is a cash flow analysis, non-cash expenses such as depreciation and deferred income taxes will not be included.

Financial Evaluation

The analysis begins with the financial evaluation as all costs and revenues in accounting records are at market values. Table 4 shows the details of cattle raising costs in local currency units (LC) per head per year. As shown, it costs LC 320 per head per year to maintain brood cows and bulls and LC 2,385 per head per year for growing animals. Of this latter value LC 2000 is the cost of feed. For convenience an exchange rate of LC 10 = US $1 has been used. Costs are shown as domestic costs, foreign exchange, and subsidies and taxes. This division is in preparation for the economic evaluation that follows. Domestic costs are the costs of local or national inputs. Foreign exchange signifies the local currency cost of foreign exchange to purchase imported inputs. For example, feed costs of LC 2000 represent domestic costs of LC 600 and foreign exchange costs of LC 1400. This indicates a high reliance on imported feedstuffs.

Table 4
Production Costs
Cattle Raising

In local currency units per head per year		Market Costs
Cow/Bull Maintenance		
Hay/Pasture		LC 250
Veterinary/health		15
Domestic costs	10	
Foreign exchange costs	5	
Labor (8 hours) @ LC 2.5/hr.		20
Building and equipment maintenance		25
Domestic costs	15	
Foreign exchange costs	10	
Miscellaneous		10
Total production costs		LC 320
Domestic costs	LC 305	
Foreign exchange costs	LC 15	
Growing Animals		
Pasture		LC 150
Feed Costs		2,000
Domestic costs	600	
Foreign exchange costs	1,400	
Veterinary/health		10
Domestic costs	5	
Foreign exchange costs	5	
Labor (2 hours) @ 2.5/hr.		5
Building and equipment maintenance		15
Domestic costs	10	
Foreign exchange costs	5	
Miscellaneous		5
Taxes		200
Total production costs		LC 2,385
Domestic costs	LC 775	
Foreign exchange costs	LC 1,410	
Subsidies and taxes	LC 200	
NOTE:		
Average feed consumptions per animal per year	1.33 tons	
Cost per ton of feed		LC 1,500
Domestic costs	450	
Foreign exchange costs	1,050	
Average feed costs per animal per year		LC 2,000
Domestic costs	600	
Foreign exchange costs	1,400	

Table 5 provides the costs for slaughtering, breaking, and boning —most of which is the cost for the live animal. Note that the wage rate of LC 10 per hour for semiskilled labor is considerably higher that the LC 2.5 per hour for unskilled rural labor in cattle raising. The higher wage is also applicable in meat processing. Also note that slaughtering is evidently financially unprofitable at market prices and costs. The government is providing an LC 800 per head subsidy to the slaughterhouse.

Table 5

Production Costs
Slaughtening, Breaking and Boning

Per Head		Market Costs
Raw Material Costs		
Price per ton - live weight		LC 18,000
Per 450 kg animal		LC 8,100
Operating costs		
Labor (6 hours) @ LC 10/hr		60
Other operating costs		1,320
Domestic costs	920	
Foreign exchange costs	400	
Subsidy		(800)
Taxes		500
Total Production Costs		LC 9,180

Meat processing costs for both the current and alternative technology are shown in Table 6. Both sets of costs refer to the meat product "formulas"

Table 6

Production Costs
Meat Processing

Current Technology (all meat)		
Per Head Slaughtered		**Market Costs**
Raw materials		
136 kg manufacturing meat @17/kg		LC 2,312
35 kg fat @ 4/kg		140
265 kg other @ 1.5/kg		398
436 kg		LC 2,850
Operating Costs		720
Labor	40	
Other domestic costs	505	
Foreign exchange costs	75	
Taxes	100	
Subsidy		(100)
Total production costs		LC 3,470
Cost per kilo		LC 7.9

Alternative Technology (with isolated soy protein)		
Per Head Slaughtered		**Market Costs**
Raw materials		
136 kg manufacturing meat @17/kg		LC 2,312
50.7 kg fat @ 4/kg		203
6.9 kg isolated soy protein @24/kg		166
20.5 kg water		–
335.9 kg other @ 1.5/kg		504
550.0 kg		LC 3,185
Operating Costs[1]		908
labor	50	
Other domestic costs	637	
Foreign exchange costs	95	
Taxes	126	
Total production costs		LC 4,093
Cost per kilo		LC 7.4

[1]Same cost per kilo as current technology.

shown in Table 3. The current technology requires a subsidy of LC 100 while the alternative technology does not require a subsidy. While operating costs per kilo of output are the same for both technologies, total cost per kilo is less using the alternative technology-even without the subsidy.

Economic Evaluation

All price distortions must be eliminated from the economic evaluation to determine the profitability of the alternatives to the economy. As described earlier, these price distortions have the effect of subsidies and taxes.

For example, if a company employs a laborer based on a minimum wage which is more than the value that the laborer produces, then the minimum wage includes a "subsidy" to the laborer "paid" by an equivalent "tax" on the company. If a company sells its product at a controlled price less than that which the company would receive without price controls, the price includes a "subsidy" to consumers "paid" by an equivalent "tax" on the company. If a government administers an overvalued exchange rate, then a unit of foreign exchange (for example a US dollar) costs less in local currency units (per dollar) than if the exchange rate were free. Importers are thereby "subsidized" and exporters "taxed."

The economic evaluation therefore must use undistorted prices and costs. Undistorted prices are known as shadow prices, and are determined by opportunity costs. Economic evaluation is also known as social cost-benefit analysis.

Domestic inputs and products should be valued at their opportunity costs. On the other hand, inputs and products that are, or could be imported and exported (tradeables) should be valued at world market prices.

Table 7 gives the undistorted wage for semiskilled labor and the undistorted exchange rate for foreign exchange. The ratio of the undistorted cost to the market cost provides the adjustment factor to be used later in the analysis.

Table 7

Market Costs and Undistorted Costs
(local currency units)

	Market Costs	Undistorted Costs	Adjustment Factor
Semi-skilled labor	LC 10/hr.	LC 8/hr.	0.8
Foreign exchange	LC 10/$	LC 12/$	1.2
Taxes and subsidies	Various	Zero	0

Table 8 reiterates the costs of cattle raising presented in Table 4. Tables 9 and 10 are extensions of Tables 5 and 6 respectively. In Tables 9 and 10 imported inputs are revalued at the real exchange rate of LC 12 = US $1 in place of the controlled exchange rate (used in Tables 4, 5 and 6) of LC 10 = US $1. The cost of labor in Tables 9 and 10 is devalued from the minimum wage by the adjustment factor of 0.8 indicating that the minimum wage overvalued labor's productivity. Subsidies and taxes are eliminated.

In Table 8 cattle raising is shown to be slightly more costly economically than financially because of the higher local currency value of imported feedstuffs even when taxes are eliminated.

Table 8

Economic Evaluation
Production Costs
Cattle Raising

In local currency units per head per year	Market Costs	Adjustment Factor	Undistorted Costs
Cow/Bull Maintenance	LC 320		LC 323
Domestic cost	305	1.00	305
Foreign exchange costs	15	1.20	18
Taxes	0	0	0
Growing Animals	LC 2,385		LC 2,467
Domestic cost	775	1.00	775
Foreign exchange costs	1,410	1.20	1,692
Subsidies and taxes	200	0	0

Slaughtering is less costly economically because of the lower raw material cost as shown in Table 9. The domestic cost of live cattle is kept artificially high by limiting competition from lower cost imported cattle even when the imports are valued at the real exchange rate. Without the distortion caused by limiting imports, live cattle would be worth LC 17,000 per metric ton, not LC 18,000.

Table 9

Economic Evaluation
Production Costs
Slaughtering, Breaking, and Boning

Per Head		Market Costs	Adjustment Factor		Undistorted Costs
Raw Materials					
Price per ton - live weight		18,000	—		17,000[1]
Per 450 kg animal		LC 8,100			LC 7,650
Operating Costs					
Labor (6 hours)		60	0.8		48
Other operating costs		1,320			1,400
Domestic costs	920		1.0	920	
Foreign exchange costs	400		1.2	480	
Subsidy		(800)	0		0
Taxes		500	0		0
Total Production Costs		LC 9,180			LC 9,098

[1] U.S. $55.00/cwt   FAS = Cost in foreign market
   + 8.45   Ocean freight to import
U.S. $63.45/cwt   CIF local port
U.S. $1,395/metric ton CIF
LC 16,750/metric ton CIF @ ER = 12
        250 inland freight: port to abattoir
LC 17,000/T FOB abattoir

Imports would more probably be frozen, boxed meat not live animals. The latter is used here for simplicity to show the economic effect when lower cost imported raw materials are permitted.

Note that this analysis does not show any tariffs on imports. Tariffs are a cost in the financial evaluation. Since they represent a transfer of money to the government, they are not a cost and would be eliminated in the economic evaluation.

Table 10 shows little difference between costs at market and undistorted costs. Again, the unit cost of processed meat is lower using a mixture of isolated soy protein and meat than when using meat alone.

Table 10

Economic Evaluation
Production Costs
Meat Processing

	Current Technology (all meat)		
Per Head Slaughtered	Market Costs	Adjustment Factor	Social Costs
Raw Materials			
136 kg manufacturing meat @ 17/kg	2,312		
35 kg fat @ 4/kg	140		
265 kg other @ 1.5/kg	398		
436 kg	2,850	1.0	2,850
Operating Costs	720		627
Labor	40	0.8	32
Other domestic costs	505	1.0	505
Foreign exchange costs	75	1.2	90
Taxes	100	0	0
Subsidy	(100)	0	0
Total Production Costs	LC 3,470		LC 3,477
Cost per Kilo	LC 7.9		LC 7.9

	Alternative Technology (with isolated soy protein)		
Per Head Slaughtered	Market Costs	Adjustment Factor	Social Costs
Raw Materials			
136 kg manufacturing meat @ 17/kg	2,312	1.0	2,312
50.7 kg fat @ 4/kg	203	1.0	203
6.9 kg isolated soy protein[1] @ 24/kg	166	1.2	199
20.5 kg water	—		—
335.9 kg other @ 1.5/kg	504	1.0	504
550.0 kg	3,185		3,218
Operating Costs[2]	908		791
Labor	50	0.8	40
Other domestic costs	637	1.0	637
Foreign exchange costs	95	1.2	114
Taxes	126	0	0
Total Production Costs	LC 4,093		LC 4,009
Cost per Kilo	LC 7.4		LC 7.3

[1] Imported raw material.
[2] Same cost per kilo as current method.

206

Cash Flows

Now that both financial and economic costs are known, the time required to increase the herd size 20 percent can be shown. Costs and cash flows over time can also be shown. At this point one must first reconcile with inflation.

Standard methodology would suggest doing the analysis in current prices. This both simplifies the calculations and prevents mistaken assumptions of future inflation rates. Only the interest rate, used for discounting in the last section, needs to be adjusted for inflation and be stated in real terms.

Also, one assumes that the changes in the quantities of inputs consumed and products produced are not sufficiently great to cause changes in prices. If prices are affected the analysis is not invalidated, but the feedback effects on prices must be calculated and the valuation modified accordingly.

Technical Coefficients

The higher output of increasing the herd size 20 percent is not realized for five years as shown in Table 11. At the end of the first year 20 cows, which otherwise would have been sold, are placed in the brood herd. The number of head of cattle slaughtered in the first year is, therefore, 20 less than the previous year.

Thereafter 16 additional calves are produced each year—four brood cows, four cows for market, and eight steers for market. Animals from birth to three years old increase annually from 240 to 256 to 272 to 288. (See Summary-bottom of Table 11). It is not until year 5 that market animals slaughtered increase from 59 to 71 and cull cows slaughtered increase from 20 to 24 thus, finally, realizing the increased output.

Figure 2 shows the output of food — retail cuts plus processed meats —for both alternatives.

Figure 2

Output Retail Cuts Plus Processed Meats
(metric tons per year)

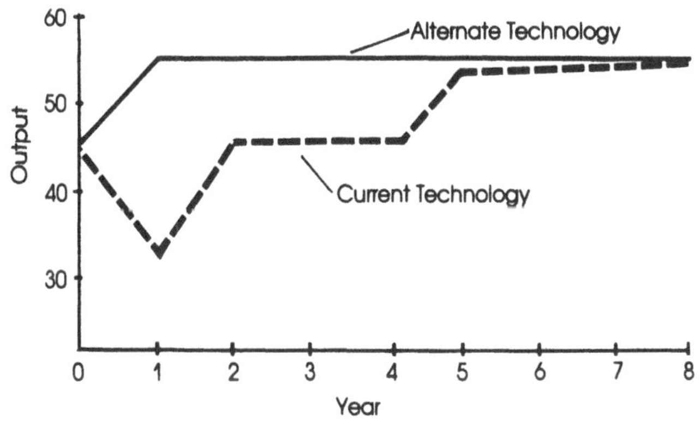

207

Table 11

Technical Coefficients
Increasing Herd Size Twenty Percent

Year:	0	1	2	3	4	5
Brood cows	100	100	120	120	120	120
Bulls	3	3	4	4	4	4
Calves produced	80	80	96	96	96	96
Cows for brood herd						
0-1 year	20	20	24	24	24	24
1-2 years	20	20	20	24	24	24
2-3 years	20	20	20	20	24	24
Bulls for brood herd						
0-1 year	1	1	1	1	1	1
1-2 years	1	1	1	1	1	1
2-3 years	1	1	1	1	1	1
Cows for market herd						
0-1 year	20	20	24	24	24	24
1-2 years	20	20	20	24	24	24
2-3 years	20	20	20	20	24	24
Steers for market						
0-1 year	39	39	47	47	47	47
1-2 years	39	39	39	47	47	47
2-3 years	39	39	39	39	47	47
Cows for slaughter	20	—	20	20	20	24
Steers for slaughter	39	39	39	39	39	47
Cull cows for slaughter	20	20	20	20	20	24
Bulls retired	1	—	1	1	1	1
Summary						
Brood cows and bulls	103	103	124	124	124	124
Animals 0-3 years	240	240	256	272	288	288
Market animals slaughtered	59	39	59	59	59	71
Cull cows slaughtered	20	20	20	20	20	24
Head Slaughtered	79	59	79	79	79	95

Financial Evaluation

Revenues and costs, at market prices, are now matched over time in Table 12. The number of animals to increase the herd size 20 percent is derived from Table 11, the price per ton from Table 5, and the costs per head from Table 4. An investment of LC 300 per additional head is required in years 1, 2, and 3. Net cash flow drops to negative in years 1 to 4 as the cost of more animals is not matched by increased revenues.

Table 12
Financial Evaluation Cash Flow
Cattle Raising

Units Per Year						
INCREASING HERD SIZE TWENTY PERCENT	0	1	2	3	4	5
REVENUES						
Head Slaughtered[1]	79	59	79	79	79	95
Total live weight - tons	35.6	26.6	35.6	35.6	35.6	42.8
Price per ton	18,000	18,000	18,000	18,000	18,000	18,000
Total Revenue	LC 640,800	478,800	640,800	640,800	640,800	770,582
PRODUCTION COSTS Brood Cows and Bulls	103	103	124	124	124	124
Cost Per Head	320	320	320	320	320	320
Cost: Brood Herd	32,960	32,960	39,680	39,680	39,680	39,680
Animals 0-3 Years	240	240	256	272	288	288
Cost Per Head	2,385	2,385	2,385	2,385	2,385	2,385
Cost: Growing Animals	572,400	572,400	610,560	648,720	686,800	686,800
TOTAL COST	LC 605,360	605,360	650,240	688,400	726,480	726,480
INVESTMENT[2]	—	11,100	4,800	4,800	—	—
NET CASH FLOW	LC 35,440	(137,660)	(14,240)	(52,400)	(85,680)	44,102

[1] At 450 kg/head
[2] LC 300/additional head

MAINTAINING SAME HERD SIZE						
NET CASH FLOW	LC 35,440	35,440	35,440	35,440	35,440	35,440

209

Alternatively, the net cash flow from a herd maintained at the same size is shown at the bottom of Table 12.

Net cash flows from the slaughtering operation are shown in Table 13. Quantities are derived from Table 2 and costs from Table 5. An investment of LC 500 per additional head is required in year 4 to accommodate the additional production in year 5.

Table 13

Financial Evaluation Cash Flow
Slaughtering, Breaking, and Boning

Per Head

REVENUES

136 kg retail cuts @ 50/kg	LC 6,800
136 kg manufacturing meat @ 17/kg	2,312
35 kg fat @ 4/kg	140
143 kg hide, shank, blood and offal @ 2/kg	286
Total Revenues	9,538
PRODUCTION COSTS	9,180
NET CASH FLOW PER HEAD	LC 358

Per Year

INCREASING HERD SIZE TWENTY PERCENT	0	1	2	3	4	5
Head Slaughtered	79	59	79	79	79	95
Net Cash Flow Per Head	LC 358	358	358	358	358	358
Operating Cash Flow	LC 28,282	21,122	28,282	28,282	20,282	34,010
Investment[1]	—	—	—	—	8,000	—
Net Cash Flow	LC 28,282	21,122	28,282	28,282	28,282	34,010
MAINTAINING SAME HERD SIZE						
Head Slaughtered	79	79	79	79	79	79
Net Cash Flow Per Head	LC 358	358	358	358	358	358
NET CASH FLOW	LC 28,282	28,282	28,282	28,282	28,282	28,282

[1]LC 500/additional head

Net cash flow for meat processing is shown in Table 14 for the current technology and in Table 15 for the alternative technology of processed meats incorporating isolated soy protein. Quantities and costs are derived from Table 6. An investment of LC 3200 is required in year 4 with the current all meat technology to increase output in year 5. With the alternative technology the increased output of processed meat products with isolated soy protein can be realized in year 1 with an investment of only LC 1,600. This is particularly significant as increased revenues more than offset increased costs immediately.

Table 14

Financial Evaluation
Cash Flow
Meat Processing

Current Technology
(all meat)

Per Head

REVENUES	
Kilos sold	436
Price per kilo	LC 8.0[1]
Total Revenues	LC 3,488
PRODUCTION COSTS	LC 3,470
NET CASH FLOW PER HEAD	LC 18

Per Year

INCREASING HERD SIZE TWENTY PERCENT	0	1	2	3	4	5
Head Slaughtered	79	59	79	79	79	95
Net Cash Flow Per Head	18	18	18	18	18	18
Operating Cash Flow	LC 1,422	1,062	1,422	1,422	1,422	1,710
Investment[2]	—	—	—	—	3,200	—
NET CASH FLOW	LC 1,422	1,062	1,422	1,422	(1,778)	1,710

[1] Controlled price
[2] LC 200/additional head

Table 15

Financial Evaluation
Cash Flow
Meat Processing

Alternative Technology
(with isolated soy protein)

Per Head

REVENUES
Kilos sold	550
Price per kilo	LC 8.0[1]

Total Revenues	LC 4,400
PRODUCTION COSTS	LC 4,093
NET CASH FLOW per head	LC 307

Per Year

USING ISOLATED SOY PROTEIN	0	1	2	3	4	5
Head Slaughtered	79	79	79	79	79	79
Net Cash Flow Per Head	18	307	307	307	307	307
Operating Cash Flow	LC 1,422	24,253	24,253	24,253	24,253	24,253
Investment	—	1,600				
NET CASH FLOW	LC 1,422	22,653	24,253	24,253	24,253	24,253

[1]Controlled price

Economic Evaluation

The above analysis is now repeated using undistorted prices and costs with some very enlightening insights.

First, Table 16 shows that cattle raising is economically unprofitable even without increasing the herd size. Only the high domestic price of live cattle makes cattle raising financially profitable. The number of animals is derived in Table 11, the price per ton from Table 9 and the costs per head in Table. 8.

212

Table 16

Financial Evaluation
Cash Flow
Cattle Raising

INCREASING HERD SIZE TWENTY PERCENT	0	1	2	3	4	5
REVENUES						
Head Slaughtered	79	59	79	79	79	95
Total live weight - tons	35.6	26.6	35.6	35.6	35.6	42.8
Price per ton	17,000	17,000	17,000	17,000	17,000	17,000
Total Revenue	LC 605,200	452,200	605,200	605,200	605,200	727,600
PRODUCTION COSTS						
Brood Cows and Bulls	103	103	124	124	124	124
Cost Per Head	323	323	323	323	323	323
Cost: Brood Herd	33,269	33,269	40,052	40,052	40,052	40,052
Animals 0-3 Years	240	240	256	272	288	288
Cost Per Head	2,467	2,467	2,467	2,467	2,467	2,467
Cost: Growing Animals	592,080	592,080	631,552	671,024	710,496	710,496
TOTAL COST	LC 625,349	625,349	671,604	711,076	750,548	750,548
INVESTMENT[1]	—	11,100	4,800	4,800	—	—
NET CASH FLOW	LC (20,149)	(184,249)	(71,204)	(110,676)	(145,348)	(22,948)

[1] LC 300/additional head

MAINTAINING SAME HERD SIZE						
NET CASH FLOW	LC (20,149)	(20,149)	(20,149)	(20,149)	(20,149)	(20,149)

Net cash flows from the slaughtering operation are shown in Table 17. The prices received for sales are assumed to be undistorted and therefore the same as Table 13. Production costs came from Table 9. Slaughtering is more profitable economically than financially because of the lower cost for live cattle.

Table 17

Financial Evaluation
Cash Flow
Slaughtering, Breaking, and Boning

Per Head

REVENUES	9,538
PRODUCTION COSTS	9,098
NET CASH FLOW per head	LC 440

Per Year

INCREASING HERD SIZE TWENTY PERCENT	0	1	2	3	4	5
Head Slaughtered	79	59	79	79	79	95
Net Cash Flow Per Head	440	440	440	440	440	440
Operating Cash Flow	LC 34,760	25,960	34,760	34,760	34,760	41,800
Investment[1]	—	—	—	—	8,000	—
NET CASH FLOW	LC 34,760	25,960	34,760	34,760	26,760	41,800

MAINTAINING SAME HERD SIZE						
Head Slaughtered	79	79	79	79	79	79
Net Cash Flow Per Head	440	440	440	440	440	440
NET CASH FLOW	LC 34,760	34,760	34,760	34,760	34,760	34,760

[1]LC 500/additional head

214

Table 18

Financial Evaluation
Cash Flow
Meat Processing

Current Technology
(all meat)

Per Head

REVENUES
Kilos sold	436
Price per kilo	LC 8.2[1]

Total Revenues .... LC 3,575

PRODUCTION COSTS .... LC 3,477

NET CASH FLOW per head .... LC 98

Per Year

INCREASING HERD SIZE TWENTY PERCENT	0	1	2	3	4	5
Head Slaughtered	79	59	79	79	79	95
Net Cash Flow Per Head	LC 98	98	98	98	98	98
Operating Cash Flow	LC 7,742	5,782	7,742	7,742	7,742	9,310
Investment[2]	—	—	—	—	3,200	—
NET CASH FLOW	LC 7,742	5,782	7,742	7,742	4,542	9,310

[1]Controlled price was LC 8.0/kg
[2]LC 200/additional head

Net cash flows for meat processing are shown in Table 18 for the current technology and in Table 19 for the alternative technology using isolated soy protein. Quantities and costs are derived from Table 10. This operation is more profitable economically than financially as the controlled price for processed meat products does not reflect their full value. Again the alternative technology with isolated soy protein is significantly more profitable than the current technology because of the lower production cost per kilo.

Table 19

Financial Evaluation
Cash Flow
Meat Processing

Alternative Technology
(with isolated soy protein)

<u>Per Head</u>

REVENUES

Kilos sold	550
Price per kilo	LC 8.2[1]
Total Revenues	LC 4,510
PRODUCTION COSTS	LC 4,009
Net Cash Flow per head	LC 501

<u>Per Year</u>

USING ISOLATED SOY PROTEIN	0	1	2	3	4	5
Head Slaughtered	79	79	79	79	79	79
Net Cash Flow Per Head	LC 98	501	501	501	501	501
Operating Cash Flow	LC 7,742	39,579	39,579	39,579	39,579	39,579
Investment	—	1,600				
NET CASH FLOW	LC 7,742	37,979	39,579	39,579	39,579	39,579

[1]Controlled price was LC 8.0/kg

Quantifying Profitability

The relative profitability of the two technologies is measured—both financially and economically—by adding and discounting the cash flow from the operations to determine the net present value.

The financial evaluation is shown in Table 20. With the current technology, the herd size, and the output of cattle raising, slaughtering, and meat processing

216

are all increased 20 percent. The cash flows came from Tables 12, 13, and 14. The net present value over eight years is LC 118,533. It does not turn positive until year 6.

Table 20
Comparison of Net Present Values
Financial Evaluation

INCREASING HERD SIZE TWENTY PERCENT

Net Cash Flow		0	1	2	3	4	5	6	7	8
Cattle Raising (T12)	LC	35,440	(137,660)	(14,240)	(52,400)	(85,680)	44,102			
Slaughtering, Breaking and Boning (T13)	LC	28,282	21,122	28,282	28,282	20,282	34,010			
Meat Processing: Current Technology (T14)	LC	1,422	1,062	1,422	1,422	(1,778)	1,710			
Total	LC	65,144	(115,476)	15,464	(22,696)	(67,176)	79,822	79,822	79,822	79,822
Discount Factor @ 6%		1.00	0.94	0.89	0.84	0.79	0.75	0.71	0.67	0.63
Discounted Flow	LC	65,144	(108,547)	13,763	(19,065)	(53,069)	59,866	56,673	53,480	50,287
Net Present Value	LC	118,533								

USING ISOLATED SOY PROTEIN

Net Cash Flow		0	1	2	3	4	5	6	7	8
Cattle Raising (T12)	LC	35,440	35,440	35,440	35,440	35,440	35,440			
Slaughtering, Breaking and Boning (T13)	LC	28,282	28,282	28,282	28,282	28,282	28,282			
Meat Processing: Alternative Technology (T15)	LC	1,422	22,653	24,253	24,253	24,253	24,253			
Total	LC	65,144	86,357	87,975	87,975	87,975	87,975	87,975	87,975	87,975
Discount Factor @ 6%		1.00	0.94	0.89	0.84	0.79	0.75	0.71	0.67	0.63
Discounted Flow	LC	65,144	81,176	78,297	73,893	69,500	65,981	62,462	58,943	55,424
Net Present Value	LC	610,820								

With the alternative technology, the herd size and the output of cattle raising and slaughtering are not increased, but the output of meat processing is increased 26.1 percent through the use of isolated soy protein. The cash flows come from Tables 12, 13, and 15. The net present value is LC 610,820 over eight years. It is continually positive and exceeds that for the alternative for all points in time. This is shown in Figure 3.

Again, the economic analysis provides some real insights as shown in Table 21. Increasing output 20 percent by increasing cattle output is economically unprofitable in four of the first five years and has a negative net present value of LC-216,184 over eight years.

Table 21
Comparison of Net Present Values
Economic Evaluation

INCREASING HERD SIZE TWENTY PERCENT

Net Cash Flow		0	1	2	3	4	5	6	7	8
Cattle Raising (T16)	LC	(20,149)	(184,249)	(71,204)	(110,676)	(145,348)	(22,948)			
Slaughtering, Breaking and Boning (T17)	LC	34,760	25,960	34,760	34,760	26,760	41,800			
Meat Processing: Current Technology (T18)	LC	7,742	5,782	7,742	7,742	4,542	9,310			
Total	LC	22,353	(152,507)	(28,702)	(68,174)	(114,046)	28,162	28,162	28,162	28,162
Discount Factor @ 6%		1.00	0.94	0.89	0.84	0.79	0.75	0.71	0.67	0.63
Discounted Flow	LC	22,353	(143,356)	(25,545)	(57,266)	(90,096)	21,121	19,995	18,868	17,742
Net Present Value	LC	(216,184)								

USING ISOLATED SOY PROTEIN

Net Cash Flow		0	1	2	3	4	5	6	7	8
Cattle Raising (T16)	LC	(20,149)	(20,149)	(20,149)	(20,149)	(20,149)	(20,149)			
Slaughtering, Breaking and Boning (T17)	LC	34,760	34,760	34,760	34,760	34,760	34,760			
Meat Processing: Alternative Technology (T19)	LC	7,742	37,979	39,579	39,579	39,579	39,579			
Total	LC	22,353	52,590	54,190	54,190	54,190	54,190	54,190	54,190	54,190
Discount Factor @ 6%		1.00	0.94	0.89	0.84	0.79	0.75	0.71	0.67	0.63
Discounted Flow	LC	22,353	44,434	48,229	45,519	42,810	40,642	38,475	36,307	34,135
Net Present Value	LC	357,908								

Alternatively increasing total output 20 percent by maintaining the same herd size and using isolated soy protein to increase processed meat production is economically profitable and shows a net present value over eight years of LC 347,908. These flows are shown in Figure 4.

218

Figure 3

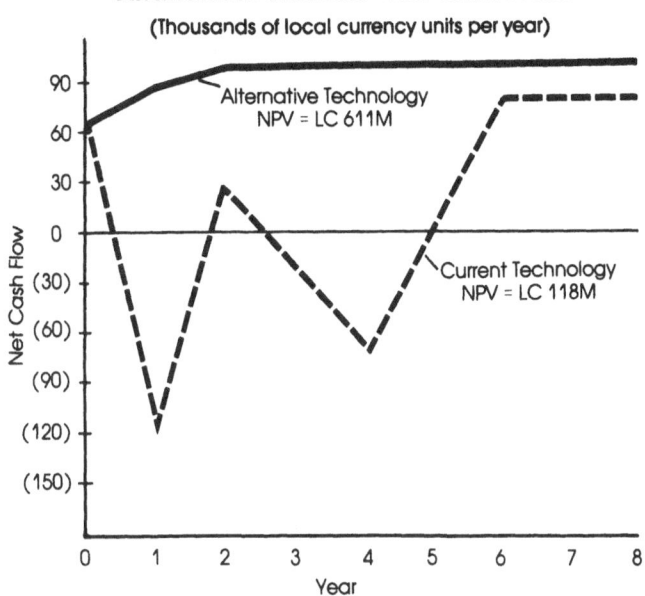

Financial Evaluation - Net Cash Flows

(Thousands of local currency units per year)

Figure 4

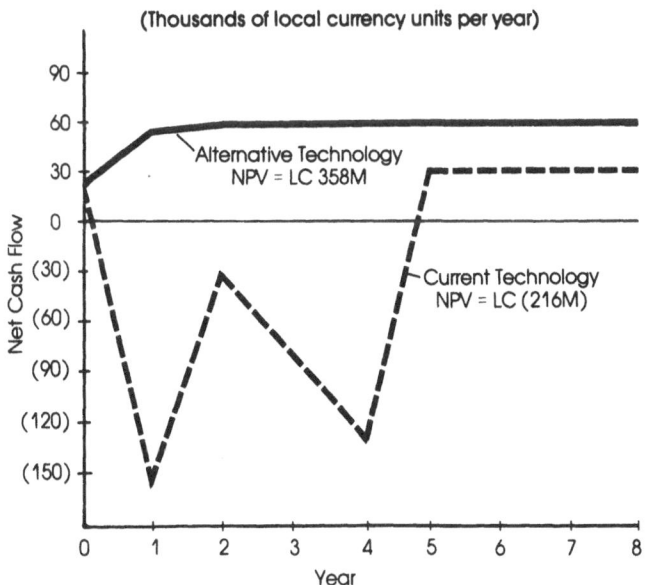

Economic Evaluation - Net Cash Flows

(Thousands of local currency units per year)

Further Considerations

Two issues do not come out in the analysis but seem relevant. These are transportation and storage costs and fixed versus variable costs.

The example assumes that appropriate marketing and processing institutions are in place, and that all operations from feedstuff production to final meat consumption take place within reasonable geographic proximity. This is not always the case—China being a good example.

When transportation and storage costs become important, they should be included in the analysis. For example, from Tables 1, 2 and 3, transporting one ton of isolated soy protein is equivalent to transporting five tons of 80 percent lean, boneless beef $(6.9+20.5)/6.9/0.8$ or 11 tons of carcass beef $5(307/136)$ or 16 tons (live weight) of cattle $5(450/136)$ or 142 tons of feed $16(4000/450)$. While beef requires refrigerated transportation and storage, isolated soy protein does not. The cost differences can be significant.

Second, the example does not fully differentiate between fixed and variable costs. The alternative technology, with isolated soy protein, does have lower fixed to total costs. If there is considerable variability in demand and/or in the supply of meat, the option with the lower fixed costs would permit more flexibility.

Further, isolated soy protein could be used in the short-run while long-run adjustments are taking place in the livestock industry. In fact, the increase in output and/or productivity in cattle production is quite complementary with the use of isolated soy protein as will be further demonstrated in the Appendix.

Summary and Conclusion

The purpose of this chapter is to describe the methodology to quantify the financial and economic profitability of using isolated soy protein to increase the meat supply. The analysis demonstrated that the use of this ingredient is both financially and economically profitable *per se* and more profitable than increasing cattle output by an equivalent amount. The net present values of these alternatives are:

Net Present Value

(in thousands)

Alternative	Financial Evaluation	Economic Evaluation
Increasing herd size 20%	LC 118	LC (216)
Using isolated soy protein	LC 610	LC 357

( ) indicates negative

Important steps in the analysis include:

1. Identifying the technical coeffecients that describe and balance the relationship between inputs and outputs for cattle raising, slaughtering, and meat processing.

2. Identifying prices and determining all costs paid and revenues received for a cash flow analysis of the three operations.

3. Correcting all cost and price distortions to reflect opportunity costs for domestic inputs and products and world prices for all tradeables. Eliminating all taxes and subsidies.

4. Calculating the net cash flow and net present values in both financial and economic terms for both alternatives.

5. Evaluating the profitability of each alternative individually and relatively.

The example demonstrated the major factors affecting profitability. The single largest cost throughout this example is (feed conversion and) the cost of feed.

The higher domestic market price for live cattle and the controlled price of processed meats have significant financial effects, respectively, on raising the profits of cattle producers and lowering the profits for meat processors. The overvalued exchange rate is noteworthy only where it raises the cost of imported feed ingredients. Adjusting the overvalued minimum wage rate for semi-skilled labor is of little importance. Taxes and subsidies are important *per se* but when summed have a negligible net effect.

Financially, the use of isolated soy protein in this example significantly increased profit to the system while simultaneously eliminating the need for subsidies to meat processors. Economically, isolated soy protein overcame the higher domestic cost of cattle and eliminated the need for subsidies to both slaughtering and meat processing operations. It would appear in this example that the use of isolated soy protein should be used by the meat processing industry and its imports and use be supported by government policies.

Appendix

Financial and Economic Sensitivity Analysis

From the absolute magnitude of the costs in this example, the conclusion stated that, "The single largest cost throughout this example is (feed conversion and) the cost of feed." To demonstrate the methodology of sensitivity analysis, this appendix shows the effect of changes in feed conversion and the cost of feed on profitability.

It would be more meaningful if the ranges of possible values for key technical coefficients and ranges of key costs could be stated. From combinations of those coefficients and costs the range of profitability could be calculated. Unfortunately, the ranges are too broad and the combinations too great to permit such an analysis.

Rather, it is important to realize that profitability can be greatly affected by changes in technical coefficients, prices, and government polices and improvements. Further, both increases in cattle production are not competitive with, but rather, are complementary to the use of isolated soy protein.

To demonstrate the sensitivity of these changes two examples are given below. The first example, given in part, changes feed conversion from 10:1 to 8:1 as shown in Table A1. The resulting cost calculations are given in Table A2. The second example changes both feed conversion, as above, and reduces feed costs from LC 1500 per ton to LC 1200 per ton as shown in Tables A3 and A4. Most of the reduction comes in imported feedstuffs. Full calculations are given in Tables A5 to A9.

The profitabilities of these six options are:

Net Present Value

(in thousands)

Alternative	Financial Evaluation	Economic Evaluation
Original Example		
Increasing herd size 20%	LC 118	LC (216)
Using isolated soy protein	LC 610	LC 357

Improving Feed Conversion

Increasing herd size 20%	LC 864	LC 658
Using isolated soy protein	LC 1,295	LC 1,137

Improving Feed Conversion and
Lowering Feed Cost

Increasing herd size 20%	LC 1,510	LC 1,387
Using isolated soy protein	LC 1,851	LC 1,787

(   ) indicates negative

The magnitude of the changes in the net present values confirms that profitability can be greatly affected. They also show, in this example, that improved cattle production and the use of isolated soy protein offer complementary benefits—a point that has been repeatedly observed in business experience.

This is not to imply that isolated soy protein is always the alternative of choice. What is implied and demonstrated is that, while the relevant methods are necessarily complex, good answers can be developed.

Table A1

Technical Coefficients
Cattle Raising

Improved Feed Conversion

Per 100 Cows

	Male	Female
Number of cows	100	
Number of bulls	3	
Calf crop	80%	
Cow replacement per year	20%	
Years in breeding herd	4 years	
Calves per year	40	40
Replacement per year	1	20
Available per year as market animals	39	20

Breeding stock	
Cows	100
Bulls	3
Calves 0-1 year	21
1-2 years	21
2-3 years	21

Stock for market	
0-1 years	59
1-2 years	59
2-3 years	59
Total herd	343

Market animals sold per year	59
Cull cows sold per year	20
Bulls retired per year	1
Total calves per year	80

Age at market	3 years
Weight at market	450 kg
Weight gain on feed	400 kg
Feed conversion	8:1*
Feed consumption per head 0-3 years	3,200 kg*
Average tons of feed per year	$3.2 \div 3 =$ 1.07 tons/head*

Tons (live weight) sold per year:		
Market animals	(59 head)	26.6
Cull cows	(20 head)	9.0
Total	(79 head)	35.6

Labor inputs are shown in Table 4.

*Indicates changes to Table 1.

225

Table A2

Production Costs
Cattle Raising

Improved Feed Conversion

In local currency units per head per year		Market Costs
Cow/Bull Maintenance		
Hay/Pasture		LC 250
Veterinary/health		15
Domestic costs	10	
Foreign exchange costs	5	
Labor (8 hours) @ LC 2.5/hr.		20
Building and equipment maintenance		25
Domestic costs	15	
Foreign exchange costs	10	
Miscellaneous		10
Total production costs		LC 320
Domestic costs	LC 305	
Foreign exchange costs	LC 15	
Growing Animals		
Pasture		LC 150
Feed Costs		1,605*
Domestic costs	482*	
Foreign exchange costs	1,123*	
Veterinary/health		10
Domestic costs	5	
Foreign exchange costs	5	
Labor (2 hours) @ 2.5/hr.		5
Building and equipment maintenance		15
Domestic costs	10	
Foreign exchange costs	5	
Miscellaneous		5
Taxes		200
Total production costs		LC 1,990*
Domestic costs	LC 657*	
Foreign exchange costs	LC 1,133*	
Interest and taxes	LC 200	
NOTE:		
Average feed consumptions per animal per year	1.07 tons*	
Cost per ton of feed		LC 1,500*
Domestic costs	450	
Foreign exchange costs	1,050	
Average feed costs per animal per year		LC 1,605*
Domestic costs	482*	
Foreign exchange costs	1,123*	

*Indicates changes to Table 4.

Table A3

Technical Coefficients
Cattle Raising

Improved Feed Conversion and Lower Feed Cost

Per 100 Cows

Number of cows		100	
Number of bulls		3	
Calf crop		80%	
Cow replacement per year		20%	
Years in breeding herd		4 years	
	Male		**Female**
Calves per year	40		40
Replacement per year	1		20
Available per year as market animals	39		20
Breeding stock			
Cows		100	
Bulls		3	
Calves 0-1 year		21	
1-2 years		21	
2-3 years		21	
Stock for market			
0-1 years		59	
1-2 years		59	
2-3 years		59	
Total herd		343	
Market animals sold per year		59	
Cull cows sold per year		20	
Bulls retired per year		1	
Total calves per year		80	
Age at market		3 years	
Weight at market		450 kg	
Weight gain on feed		400 kg	
Feed conversion		8:1*	
Feed consumption per head 0-3 years		3,200 kg*	
Average tons of feed per year		3.2 ÷ 3 = 1.07 tons/head*	

Tons (live weight) sold per year:

Market animals	(59 head)	26.6	
Cull cows	(20 head)	9.0	
Total	(79 head)	35.6	

Labor inputs are shown in Table 4.

*Indicates changes to Table 1.

Table A4

Production Costs
Cattle Raising

Improved Feed Conversion and Lower Feed Cost

In local currency units per head per year		Market Costs
Cow/Bull Maintenance		
Hay/Pasture		LC 250
Veterinary/health		15
Domestic costs	10	
Foreign exchange costs	5	
Labor (8 hours) @ LC 2.5/hr.		20
Building and equipment maintenance		25
Domestic costs	15	
Foreign exchange costs	10	
Miscellaneous		10
Total production costs		LC 320
Domestic costs	LC 305	
Foreign exchange costs	LC 15	
Growing Animals		
Pasture		LC 150
Feed Costs		1,284*
Domestic costs	428*	
Foreign exchange costs	856*	
Veterinary/health		10
Domestic costs	5	
Foreign exchange costs	5	
Labor (2 hours) @ 2.5/hr.		5
Building and equipment maintenance		15
Domestic costs	10	
Foreign exchange costs	5	
Miscellaneous		5
Taxes		200
Total production costs		LC 1,669*
Domestic costs	LC 603*	
Foreign exchange costs	LC 866*	
Interest and taxes	LC 200	
NOTE:		
Average feed consumptions per animal per year	1.07 tons*	
Cost per ton of feed		LC 1,200*
Domestic costs	400*	
Foreign exchange costs	800*	
Average feed costs per animal per year		LC 1,284*
Domestic costs	428*	
Foreign exchange costs	856*	

*Indicates changes to Table 4.

Table A5

Financial Evaluation
Cash Flow
Cattle Raising

Improved Feed Conversion and Lower Feed Cost

Units Per Year

INCREASING HERD SIZE TWENTY PERCENT	0	1	2	3	4	5
REVENUES						
Head Slaughtered[1]	79	59	79	79	79	95
Total live weight - tons	35.6	26.6	35.6	35.6	35.6	42.8
Price per ton	18,000	18,000	18,000	18,000	18,000	18,000
Total Revenue	LC 640,800	478,800	640,800	640,800	640,800	770,582
PRODUCTION COSTS						
Brood Cows and Bulls	103	103	124	124	124	124
Cost Per Head	320	320	320	320	320	320
Cost: Brood Herd	32,960	32,960	39,680	39,680	39,680	39,680
Animals 0-3 Years	240	240	256	272	288	288
Cost Per Head	1,669*	1,669*	1,669*	1,669*	1,669*	1,669*
Cost: Growing Animals	400,560*	400,560*	427,264*	453,960*	480,672*	480,672*
TOTAL COST	LC 433,520*	433,520*	466,944*	493,648*	520,352*	520,352*
INVESTMENT[2]	—	11,100	4,800	4,800	—	—
NET CASH FLOW	LC 207,280*	34,180*	169,056*	142,352*	120,448*	250,230*

[1] At 450 kg/head
[2] LC 300/additional head

MAINTAINING SAME HERD SIZE						
NET CASH FLOW	LC 207,280*	207,280*	207,280*	207.280*	207,280*	207,280*

*Indicates changes to Table 12.

Table A6

Comparison of Net Present Values
Financial Evaluation

Improved Feed Conversion and Lower Feed Cost

INCREASING HERD SIZE TWENTY PERCENT

Net Cash Flow		0	1	2	3	4	5	6	7	8
Cattle Raising (A5)	LC	207,280*	34,180*	169,056*	142,352*	120,448*	250,230*			
Slaughtering, Breaking and Boning (T13)	LC	28,282	21,122	28,282	28,282	20,282	34,010			
Meat Processing: Current Technology (T14)	LC	1,422	1,062	1,422	1,422	(1,778)	1,710			
Total	LC	236,984*	56,364*	198,760*	172,056*	138,952*	285,950*	285,950*	285,950*	285,950*
Discount Factor @ 6%		1.00	0.94	0.89	0.84	0.79	0.75	0.71	0.67	0.63
Discounted Flow	LC	236,984*	52,982*	176,896*	144,527*	109,772*	214,463*	203,025*	191,587*	180,149*
Net Present Value	LC	1,510,385*								

USING ISOLATED SOY PROTEIN

Net Cash Flow		0	1	2	3	4	5	6	7	8
Cattle Raising (A5)	LC	207,280*	207,280*	207,280*	207,280*	207,280*	207,280*			
Slaughtering, Breaking and Boning (T13)	LC	28,282	28,282	28,282	28,282	28,282	28,282			
Meat Processing: Alternative Technology (T15)	LC	1,422	22,653	24,253	24,253	24,253	24,253			
Total	LC	236,984*	258,215*	259,815*	259,815*	259,815*	259,815*	259,815*	259,815*	259,815*
Discount Factor @ 6%		1.00	0.94	0.89	0.84	0.79	0.75	0.71	0.67	0.63
Discounted Flow	LC	236,984*	242,722*	231,235*	218,245*	205,254*	194,861*	184,469*	174,076*	163,683*
Net Present Value	LC	1,851,529*								

*Indicates changes to Table 20.

Table A7

Economic Evaluation
Production Costs
Cattle Raising

Improved Feed Conversion and Lower Feed Cost

In local currency units per head per year	Market Costs	Adjustment Factor		Undistorted Costs
Cow/Bull Maintenance	LC 320			LC 323
Domestic cost	305	1.00	305	
Foreign exchange costs	15	1.20	18	
Taxes	0	0	0	
Growing Animals	LC 1,669*			LC 1,642*
Domestic cost	603*	1.00	603*	
Foreign exchange costs	866*	1.20	1,039*	
Interest and taxes	200	0	0	

*Indicates changes to Table 8.

Table A8

Economic Evaluation
Cash Flow
Cattle Raising

INCREASING HERD SIZE TWENTY PERCENT	Improved Feed Conversion and Lower Feed Cost					
	0	1	2	3	4	5
REVENUES						
Head Slaughtered	79	59	79	79	79	95
Total live weight - tons	35.6	26.6	35.6	35.6	35.6	42.8
Price per ton	17,000	17,000	17,000	17,000	17,000	17,000
Total Revenue	LC 605,200	452,200	605,200	605,200	605,200	727,600
PRODUCTION COSTS						
Brood Cows and Bulls	103	103	124	124	124	124
Cost Per Head	323	323	323	323	323	323
Cost: Brood Herd	33,269	33,269	40,052	40,052	40,052	40,052
Animals 0-3 Years	240	240	256	272	288	288
Cost Per Head	1,642*	1,642*	1,642*	1,642*	1,642*	1,642*
Cost: Growing Animals	394,080*	394,080*	420,352*	446,624*	472,896*	472,896*
TOTAL COST	LC 427,349*	427,349*	460,404*	486,676*	512,948*	512,948*
INVESTMENT[1]	—	11,100	4,800	4,800	—	—
NET CASH FLOW	LC 177,851*	13,751*	139,996*	113,724*	92,252*	214,652*

[1] LC 300/additional head

MAINTAINING SAME HERD SIZE						
NET CASH FLOW	LC 177,851*	177,851*	177,851*	177,851*	177,851*	177,851*

*Indicates changes to Table 16.

Table A9

Comparison of Net Present Values
Economic Evaluation

Improved Feed Conversion and Lower Feed Cost

INCREASING HERD SIZE TWENTY PERCENT

Net Cash Flow		0	1	2	3	4	5	6	7	8
Cattle Raising (A8)	LC	177,851*	13,751*	139,996*	113,724*	92,252*	214,652*			
Slaughtering, Breaking and Boning (T17)	LC	34,760	25,960	34,760	34,760	26,760	41,800			
Meat Processing: Current Technology (T18)	LC	7,742	5,782	7,742	7,742	4,542	9,310			
Total	LC	220,353*	45,493*	182,498*	156,226*	123,554*	265,762*	265,762*	265,762*	265,762*
Discount Factor @ 6%		1.00	0.94	0.89	0.84	0.79	0.75	0.71	0.67	0.63
Discounted Flow	LC	220,353*	42,763*	162,423*	131,230*	97,608*	199,322*	188,691*	178,061*	167,430*
Net Present Value	LC	1,387,881*								

USING ISOLATED SOY PROTEIN

Net Cash Flow		0	1	2	3	4	5	6	7	8
Cattle Raising (A8)	LC	177,851*	177,851*	177,851*	177,851*	177,851*	177,851*			
Slaughtering, Breaking and Boning (T17)	LC	34,760	34,760	34,760	34,760	34,760	34,760			
Meat Processing: Alternative Technology (T19)	LC	7,742	37,979	39,579	39,579	39,579	39,579			
Total	LC	220,353*	250,590*	252,190*	252,190*	252,190*	252,190*	252,190*	252,190*	252,190*
Discount Factor @ 6%		1.00	0.94	0.89	0.84	0.79	0.75	0.71	0.67	0.63
Discounted Flow	LC	220,353*	235,555*	224,449*	211,840*	199,230*	189,143*	179,055*	168,967*	158,880*
Net Present Value	LC	1,787,472*								

*Indicates changes to Table 21.

233

Bibliography

Corden, W.M., *Trade Policy and Economic Welfare*, Claredon Press, Oxford, 1974.

Gittinger, J. Price, *Economic Analysis of Agricultural Projects*, The Johns Hopkins University Press, Baltimore, 1972.

Roemer, Michael, and Joseph J. Stern, *The Appraisal of Development Projects*, Praeger, New York, 1975.

Squire, Lynn, and Herman G. van der Tak, *Economic Analysis of Projects*, The Johns Hopkins University Press, Baltimore, 1975.

Timmer, C. Peter, Walter P. Falcon, and Scott R. Pearson, *Food Policy Analysis*, The Johns Hopkins University Press, Baltimore, 1983.

United Nations Industrial Development Organization, *Guidelines Project Evaluation*, United Nations, New York, 1972.

Chapter 7

The Relative Merits of Various Protein Food Ingredients

*Walter J. Wolf**

*Walter J. Wolf is Research Chemist, Northern Regional Research Center, United States Department of Agriculture, Peoria, Illinois.

Abstract

Soy proteins appear to be the only commercial protein ingredients that will be used in the foreseeable future for both nutritional and functional purposes as replacements for traditional animal proteins. Of various soy protein forms, isolates are the most expensive, but often offer the greatest savings and best performance functionally plus, consumer acceptance.

This chapter reviews a wide variety of alternatives to traditional animal protein (meat) ingredients. Ideal protein ingredients offer easy digestibility, high quality, safety, absence of microbial contamination, and a wide range of functional properties, including desirable organoleptic properties, low cost and wide availability.

This thorough analysis considers non-traditional animal sources, such as blood proteins and fish protein concentrate, and other plant proteins such as leaves, cottonseed, sunflower, rapeseed, winged bean, coconut, peanuts and wheat gluten. Single cell proteins hold a small position in the market, primarily as health foods and flavoring ingredients.

In addition to soy protein isolate, commonly used and widely accepted protein food ingredients include egg albumin and the milk derivatives, nonfat dry milk and sodium caseinate.

Meat proteins have established standards of nutritional and functional properties, and are the measure against which most vegetable protein food ingredients are compared. The protein food ingredients surveyed here are compared on the basis of such attributes as essential amino acid content, digestibility, safety, economics, availability, production feasibility, and functional properties such as solubility, viscosity enhancement, emulsification, cohesion-adhesion, health characteristics, flavor, texture and color.

Functional properties are critical in manufacturing and in consumer acceptance of food products containing vegetable protein ingredients.

Because of increasing costs of traditional animal proteins, such as meat and milk, and because of increasing health concerns related to these traditional products, the food industry has investigated and developed a variety of alternative sources of protein. Most of the viable alternatives are plants.

Among the non-traditional animal proteins, blood proteins have limitations of color, amino acid composition and cost. Fish protein concentrate, developed and used in Japan, is expensive, subject to fluctuations in supply, contains residual ethanol from processing, and is insoluble, thereby having limited functional value.

In their limited applications, single cell proteins are very expensive. Some yeasts have excessive nucleic acid content.

Peanuts and wheat are used commercially, but are not major factors in the protein ingredient market. Peanuts are too expensive to compete with soybeans and do not have a good balance of essential amino acids. Wheat gluten serves a special market, supplying desired baking characteristics in bread items and chewy texture in meat analogs.

Other plant protein sources—leaf protein, cottonseed, sunflower, rapeseed, winged bean and coconut—remain potentially valuable sources, but near-term commercial development is uncertain. The uncertainties include questions of safety, quality, availability and economics.

Soybeans are clearly the most promising alternative source of protein food ingredients. They meet virtually all of the desired performance and economic requirements. The soybean industry is well established and a wide range of products is available in the form of flours, protein concentrates and isolates.

Isolates are the most highly refined and most expensive form, but are widely available and have the greatest versatility. They often offer the greatest savings because they can be added to foods at higher levels and still retain all their performance advantages.

Introduction

Many food products found on United States supermarket shelves today are made from a variety of ingredients including some form of protein such as egg albumin, nonfat dry milk, sodium caseinate or soy protein isolate. The protein ingredient may be used at a level where it supplies a significant part of the dietary requirements for essential amino acids and thus makes an important nutritional contribution. An example is soy protein isolate added to ground meat products and sausages. Alternatively, the protein may be added at low levels ($\sim$ 1-2%) to provide functional properties instead of nutrition. Example of the latter application are sodium caseinate as a stabilizer in whipped toppings and coffee whiteners.

Proteins used as food ingredients are derived from both animal and plant sources. Limited supplies and increasing prices of traditional animal proteins have, however, focused attention on a number of alternate protein sources including plants, microorganisms and unconventional animal proteins. The purpose of this paper is to examine the various alternate proteins that may be used as food ingredients to replace the traditional animal proteins and to compare their advantages and disadvantages, availability and costs.

Characteristics of Animal Proteins

Animal proteins have a number of properties that have resulted in their preferential use over plant proteins, particularly in developed countries such as the United States, Canada and the Western European nations. Although their usage is often guided by tradition and availability, animal proteins have established nutritional and functional properties. Factors such as health concerns and costs have made animal proteins less attractive than plant proteins in recent years.

Tradition of Usage

A long history of consumption of milk, meat and eggs in Western countries has firmly established these products as preferred protein forms in the diet. This preference is reflected in the greater consumption of animal proteins in the United States as compared to Japan which has a long tradition of consuming vegetable proteins such as rice and soybeans.

In the United States, consumption of meat, poultry and dairy products is 3-4 times that in Japan, whereas legumes and cereals are eaten in smaller quantities than in Japan. Meat occupies a leadership position in many countries as the most sought after protein source. For example, meat has a high status as a food in the United States, and meals are frequently planned around the meat dish that is selected by the housewife.

Table 1   Per Capita Consumption of Protein-Containing Foodstuffs in the United States and Japan (1981)

Foodstuff	U.S.[a] g/day/capita	Japan[b]
Meat and poultry	273.5	61.9
Fish and other seafoods	21.3	96.1
Eggs	41.8	40.4
Milk and dairy products	378.2	177.6
Fruit	205.6	106.6
Vegetables	254.2	308.8
Potatoes	98.8	47.6
Legumes	18.7	23.1
Cereals	187.8	307.5
Miso and soy sauce	—	45.7

[a] USDA (1982).

[b] Ministry of Agriculture, Forestry and Fisheries (1982).

Availability

In contrast to plant protein products such as cereals and legumes, many animal proteins (meat, milk and eggs) are more perishable. However, widespread availability of refrigerators in marketing channels and homes makes it possible to successfully market and consume perishable animal products. The long period of useage of animal protein foods has resulted in well-developed practices and regulations for their manufacture, storage, transportation and marketing to ensure quality and microbiological safety to the consumer and thus further enhance acceptance and availability of these foods. /

The United States has a large animal industry that provides most of the meat, milk and eggs consumed in this country. Although relatively expensive, animal protein foods are readily available and thus are consumed in the quantities indicated earlier (Table 1). Defatted soybean meal is an important protein source that is available in large quantities and is used as a feed ingredient that makes it possible to support the American animal industry, especially the swine and poultry segments.

Nutritional Quality

It is well known that most animal proteins have a high nutritional quality. For example, the essential amino acid pattern of whole egg protein was adopted by FAO/WHO (1965) over two decades ago as a reference protein in recognition of its high biological value. Likewise, casein has long been used as a control in feeding tests for evaluating the Protein Efficiency Ratio (PER) of proteins in the official method used in the United States (AOAC 1984). Casein is also the reference protein in specifications for the use of textured soybean proteins as meat extenders in the National School Lunch Program; meat:soy blends must have a PER at least equivalent to that of casein, i.e., 2.5 (USDA 1971).

Some animal proteins are also excellent sources of nutrients besides essential amino acids. One such protein is casein that exists as a stable complex with calcium and makes milk a valuable dietary source of calcium. Some of the plant proteins are not soluble in the presence of high concentrations of calcium, hence analogs with the functional and nutritional properties of milk have been difficult to prepare.

Functional Properties

Certain animal proteins have additional functional properties that are highly desired and which likewise have been difficult to imitate by other proteins. Egg albumen, for example, has the characteristic of thermal setting that results in the soft, fluffy texture of baked goods such as angel food cakes. Gelatin has the unusual functional property of forming heat reversible gels at low concentrations. Casein has desirable melting and stretching properties found in cheeses; to date these properties have not been successfully duplicated with other proteins. Consequently, imitation cheese products presently available use casein as an ingredient in order to more closely simulate natural cheese.

Various animal proteins have distinctive flavors ranging from meaty and eggy to the relatively bland taste of milk. The flavors of meats and eggs are developed during cooking and are important factors in consumer preferences of animal proteins over some of the plant proteins. Textural properties such as tenderness and juiciness are important characteristics of meats whereas creaminess, softness, and stretchiness are attributed to various cheeses. The organoleptic[1] properties of animal proteins, such as the tenderness of the myofibrillar proteins that give meat its texture, are highly regarded by consumers; it is therefore no coincidence that most plant protein ingredients are used in combination with animal proteins, particularly meat. Some of the plant protein-based foods are analogs of animal products and inability to duplicate animal protein characteristics is responsible for failure of many analogs in the marketplace.

---

[1] Organoleptic: being, relating or affecting qualities such as taste, color, odor and feel that stimulate the sense organs.

Limitations of Animal Proteins

Although highly preferred by many and of good nutritional quality, animal proteins have some features that limit their use. One such feature is the presence of cholesterol and saturated fatty acids that accompany animal protein sources such as meat, milk and eggs. In Japan, for example, soy milk has become very popular in recent years as a health food (Shurtleff and Aoyagi 1984); it lacks cholesterol and is low in saturated fats. In addition, soy protein has been found to be hypocholesteremic in laboratory animals including rhesus monkeys (Terpstra et al 1984).

Another negative factor in animal protein sources is lactose in milk and related products such as nonfat dry milk. Lactose causes intestinal disturbances in individuals that are deficient in lactase in their digestive tract and thus unable to digest and absorb the sugar. Worldwide, lactose intolerance is widespread among adults. It is lowest with white adults of Scandinavian and Western European ancestry (2-19% incidence) but high among other ethnic groups (Mediterranean, African and Oriental) where more than 90% of the adult population may be affected (Bayless et al. 1971). Allergy is an additional problem associated with consumption of milk protein products. Among children, the prevalence of cow's milk allergy has been estimated to be as low as 0.3% (Collins-Williams 1956) and as high as 7% (Clein 1954) depending upon the criteria used for diagnosis.

Criteria for Selection of Protein Food Ingredients

For a protein to be used as a food ingredient a number of criteria must be met (Table 2).

Table 2   Criteria for Selection of Food Protein

Criterion	Desirable characteristics
Nutritional quality	High Good balance of essential amino acids High digestibility
Safety	Free of toxic and antinutritional factors Free of pathogens
Functional properties	Wide range such as: Solubility Viscosity enhancement Elasticity Emulsification Cohesion-adhesion Pleasant or no flavor Desirable texture Light or white color
Economics	Low and stable cost
Availability	Large reliable supply
Investment by protein manufacturer	Available for use by food processors

These criteria include nutrition, safety, functional qualities, and economics; all proteins must pass certain minimum specifications to become successful in the marketplace. Some of these criteria may have an overriding influence over others. For example, the organoleptic properties of meat make it a successful protein source even if the economics are unfavorable (high prices) providing that the standard of living is sufficiently high.

Nutritional Quality

Ideally, a protein used as a food ingredient for nutritional purposes should contain the essential amino acids in the proportion required by man. Essential amino acid requirements for man have been under investigation for over 40 years and, with increasing knowledge of human nutrition, the requirements are gradually becoming better defined and recommendations for dietary intake are revised periodically. For example, FAO/WHO/UNU (1985) has recently issued a new set of patterns for essential amino acid requirements expressed as mg of amino acid/g of protein consumed. Recommendations for children 2-5 years old, children 10-12 years old, and for adults are shown in Table 3. FAO/WHO/UNU did not establish a recommended amino acid pattern for infants but published the amino acid pattern for human milk (Table 4). Tables 3 and 4 serve as references for evaluating protein sources as will be done later in discussing the nutritional value of soybean proteins.

Although many proteins, particularly those of plant origin, do not contain an ideal ratio of essential amino acids, this often is not a serious limitation because a diet usually contains a variety of protein sources such as cereals, legumes and animal proteins. By blending proteins a reasonable balance of the essential amino acids can often be obtained. For example, soy flour proteins are high in lysine and low in methionine but can be successfully blended with cereals such as corn so that the PER of the blend approaches that of casein (Bookwalter et al 1971).

The nutritive value of a protein can be estimated by calculating its chemical score. The chemical score of a protein is obtained by expressing the content of each essential amino acid as a percentage of that same amino acid found in a reference or standard pattern such as occurs in egg proteins. The essential amino acid that deviates the most from the standard is assumed to limit utilization of the protein and the percentage for the most limiting amino acid is used as the chemical score.

For proteins to be utilized by man, they must be hydrolyzed and absorbed. Undigested proteins are unavailable for absorption, hence are lost through the feces. Protein quality thus also depends upon digestibility; a protein that has a low digestibility will have a lower biological value than is indicated by the chemical score. High digestibility is therefore a desirable property of a dietary protein.

A variety of techniques has been used to evaluate protein quality (McLaughlan 1979, NAS 1974). These methods include Nitrogen Balance, Biological Value,

Table 3 Suggested Amino Acid Requirements and Composition of Soybean Proteins

Essential amino acid	Requirements[a]			Composition		
	Child, age			Defatted	Isolates	
	2-5	10-12	Adult	meal[b]	A[c]	B[d]
	mg/g protein					
Histidine	19	19	16	26	27	22
Isoleucine	28	28	13	46	49	49
Leucine	66	44	19	78	81	78
Lysine	58	44	16	64	64	63
Methionine + Cystine	25	22	17	26	26	22
Phenylalanine + Tyrosine	63	22	19	88	93	93
Threonine	34	28	9	39	37	37
Tryptophan	11	9	5	14	15	14
Valine	35.	25	13	46	47	48

[a] Amino acid requirements from FAO/WHO/UNU (1986).

[b] Cavins et al. (1972).

[c] EDI-PRO A, Product Data Sheet, Ralston Purina Company, St. Louis, MO.

[d] PRO-FAM G-900, Product Data Sheet, Grain Processing Corporation, Muscatine, IA.

**Table 4   Essential Amino Acid Composition of Human Milk
and Infant Formula Containing Soy Protein Isolate**

Essential amino acid	Human milk[a]	Infant formula[b]	
		Amio acid content	% of human milik
	mg/70 Kcal		
Histidine	26	54	208
Isoleucine	46	98	213
Leucine	93	162	174
Lysine	66	128	194
Methionine + Cystine	42	52	124
Phenylalanine + Tyrosine	72	184	256
Threonine	43	74	172
Tryptophan	17	30	175
Valine	55	94	170
Protein	1000	2000	200

[a] FAO/WHO/UNU (1985).

[b] Formulated to contain 2 g EDI-PRO A isolated soy protin/70 kcal. Amino acid values from Product Data Sheet, Ralston Purina Company, St. Louis, MO.

Utilization and PER. The rat PER assay is relatively simple to conduct, has been used extensively, and is the official method used in the United States and Canada (AOAC 1984). Nonetheless, the rat PER assay is often criticized as a poor model for humans. The essential amino acid requirements for the rat are higher than for humans, and casein, which is used as a standard protein, does not have the amino acid content corresponding to the requirements of humans. Consequently, alternatives to feeding studies have been proposed. One such proposal is the use of human amino acid requirements to develop a scoring system analogous to the chemical score (Steinke 1979).

Safety

Freedom from toxic factors and components having undesirable biological activities is a prerequisite for proteins used in foods. Toxic factors may be intrinsic or extrinsic to a given protein source. Examples of intrinsic toxic or antinutritional factors found in plant materials include protease inhibitors, hemagglutinins, cyanogens, saponins, gossypol, lathyrogens, and allergens (Liener 1980).

Toxic factors of extrinsic origin include materials formed or introduced during processing such as browning reaction products, oxidized lipids, lysinoalanine, methionine sulfone, solvent residues, fumigants, detergents and lubricants (Yannai 1980). Improper storage and processing can result in the growth of naturally occurring microorganisms (aflatoxin production in peanuts and cottonseed) or introduction of pathogenic bacteria *(Salmonella)*.

Functional Properties

Physical and chemical properties of proteins that affect properties of food systems containing them are referred to as functional properties and play an important role in selection of a protein for a given application. Examples of proteins with well-known functional properties are wheat gluten (viscoelasticity of doughs), gelatin (heat reversible gelation) and egg albumin (heat setting foams). Proteins used in foods are usually mixtures of different molecular species. Nonetheless, a given functional property attributed to a particular protein preparation may reflect properties of one of the proteins rather than those of the entire mixture. For example, when egg whites are beaten in a copper bowl, conalbumin apparently binds copper ion and the resulting protein-copper complex stabilizes the foam (McGee et al. 1984).

No single protein possesses all of the functional properties attributed to various proteins but, ideally, a protein should have a broad range of functional characteristics to make it useful in a number of food systems (Table 2). Information about the relationship between physical and chemical properties of a protein and behavior of the protein in a food is often lacking, but food technologists use heat, salts, pH, enzymes and other treatments empirically to modify or develop functional properties of proteins for specific uses (Kinsella 1982).

Sensory characteristics are important functional properties of proteins and have a direct impact on palatability of foods. Proteins added to foods should have flavors preferred by consumers or be bland so that desired flavors can be added. The presence of undesirable flavors is often a problem. Raw, defatted soybean flakes, for example, have characteristic grassy/beany and bitter flavors that are objectionable to most consumers. These undesirable flavors must be reduced in intensity or eliminated to obtain protein products satisfactory for food uses. Problems of flavor have been among the most difficult for the protein ingredients industry to solve.

Sensory properties related to texture are also significant functional properties of proteins. Tenderness, juiciness and chewiness are characteristics associated with meats and these properties are related to proteins found in these products. Another example is the stretching properties of mozzarella cheese that are imparted by casein, the major protein fraction of milk. Undesirable texture may also be a problem, such as the grittiness of fish protein concentrate that was produced experimentally in the 1960's.

Color is another sensory attribute that affects consumer acceptance of foods. Leaf proteins, for example, are green and have not been successful for food uses. White powders that do not introduce objectionable colors when incorporated into foods are ideal characteristics of protein ingredients.

Cost

Compared to lipids and carbohydrates, proteins are the most expensive ingredient used in processed foods. The ideal protein is low in price relative to other proteins so that there is an economic incentive for food processors to use it. Prices should also be stable so that food manufacturers are not subjected to wide, unpredictable fluctuations in ingredient costs. Protein ingredients are interchangeable in some products, hence high prices may result in substitution if a given protein rises sharply relative to other proteins.

Table 5 compares production figures and prices of several proteins used in processed foods. On a price per pound of protein basis, milk proteins range from $1.07 to $2.60, egg white costs $1.73 to $2.24, soy proteins vary from $0.30 to $1.17 and wheat gluten sells at $0.80 to $0.85. Nonfat dry milk illustrates how food processors responded to high prices for ingredients in the past decade. As a result of rapid increases in prices during the 1970's, bakers switched from nonfat dry milk to milk replacers based on blends of milk whey and soy flour or soy protein isolates. These markets for nonfat dry milk likely are lost for good unless prices drop dramatically relative to the replacers. There is a large supply of low cost whey that must be disposed of and bakers have learned how to adapt their processes to use of the replacers, hence there is additional reluctance to switch back to the more expensive all-milk product.

Table 5   U.S. Production (1982) and Selling Prices of
Proteins Used in Processed Foods

Protein source	Protein content %	Production million lb.	Price Per lb. $	Price Per lb. of protein $
Milk				
Nonfat dry milk	37	1,400	0.95-0.96	2.57-2.60
Casein, acid	95 ⎫	177[a]	1.05-1.10	1.11-1.16
Sodium caseinate	90 ⎭		1.45-1.50	1.61-1.67
Whey	14	791	0.15	1.07
Whey protein concentrate	34	—	0.44-0.48	1.29-1.41
Eggs				
Egg white, frozen	11	55	0.19	1.73
Egg white, dry	80	22	1.79	2.24
Blood				
Plasma, spray-dried		—	2.00	
Peanuts				
Defatted grits	57	—	0.80	1.40
Defatted flours	57	—	0.90	1.58
Soybeans				
Defatted flours and grits	50	350	0.15	0.30
Textured flours	50	95	0.30	0.60
Concentrates	70	80	0.43-0.65	0.61-0.93
Textured concentrates	70	8	0.52	0.74
Isolates, powdered	90	90	1.05	1.17
Wheat				
Gluten	75-80	50[b]	0.64	0.80-0.85
Yeast				
Torula yeast	50	—	1.20	2.40

[a] All imported, none produced domestically.

[b] About an equal amount is imported.

Availability

A factor that has a direct effect on price of protein ingredients is the available supply which is an important consideration in the selection of ingredients by food manufacturers. Proteins available in limited supply are subject to large changes in prices if the supply or demand changes suddenly. For example, casein is produced in comparatively fixed amounts in countries such as New Zealand and France. If milk production decreases abruptly because of a drought, casein prices rise very quickly to ration the available supply through the marketplace. Availability of casein is also determined by milk pricing policies of producing countries. For example, there is no domestic casein production in the United States; the support prices for milk make casein manufacturing prohibitively expensive. In 1981 it was estimated that the price of casein would have to be at least \$2.65/lb in order to induce domestic casein production (USDA 1981). The European Community countries subsidize casein production in order to compete successfully in world markets.

Vegetable proteins, such as those of soybean and wheat gluten, are in a better position from the standpoint of availability than some of the animal proteins because of the large potential supply of starting materials—soybeans and wheat. The bulk of soybean protein is used for animal feeds in the form of defatted meal; only about 2-3% of the total United States soybean crop is used domestically as a direct source of protein (as opposed to the indirect use as an animal feed for conversion to animal protein). Conversion of defatted soybean flakes to edible grade protein products could likely increase several fold without serious effects on the availability of defatted soybean meal for feeds.

Closely tied to availability is the ability of the industry to expand to meet increasing needs and thus avoid periodic shortages. In today's technological society it is also important that more than one source be available to supply a given ingredient in order to guarantee an uninterrupted supply. For example, when the sole supplier of cottonseed flour discontinued manufacture, users had no recourse except to stop producing their products or reformulate. Consequently, reintroduction of cottonseed flour into the marketplace will be extremely difficult in the future.

Ability to Use Capital Investment of Food Ingredient Processors

Availability of protein ingredients on a well established commercial scale enables food companies to use the capital investment and experience of the ingredient manufacturers to supply an ingredient rather than requiring the food companies to do their own development work and to build their own facilities. Food ingredient manufacturers often do development work to demonstrate utility and compatibility of a protein ingredient with various food systems. It is desirable that protein ingredients be technically capable of being used in existing food processing equipment. Ideally, the ingredient manufacturers should also conduct programs to educate consumers about the merits of new protein sources. Likewise, it is preferable that ingredient manufacturers are involved in the development of regulations that permit use of protein ingredients when they can provide a quality food product with an economic advantage to the consumer.

Alternate Protein Food Ingredients—
Merits and Limitations

Because of increasing costs of traditional animal proteins, such as meat and milk, a variety of alternative sources of protein have been developed to varying degrees. Although a few non-traditional animal proteins have been examined, most of the alternate protein sources are plants.

Blood Proteins

An animal source of protein long neglected and which is a waste disposal problem in the meat industry is blood (see reviews by Ranken 1977 and Wismer-Pedersen 1978). Blood contains 18% protein (approximately equal to protein content of lean meat) and is rich in heme iron, a form that has a high bioavailability. In countries with large animal slaughtering industries, there are substantial quantities of blood proteins available. Wismer-Pedersen (1979) estimated that in 1977 the blood from the hogs slaughtered in the EEC countries could have supplied an additional 40,000 tons of edible protein. Whole blood is added to meat products such as sausages and such use involves adding a natural product that already occurs in meat (blood content is estimated to be 0.3% of fresh beef; Warriss and Rhodes 1977). Blood may be fractionated to remove the red blood cells but this also removes almost 70% of the protein including the heme iron. Blood is collected to varying degrees in many European countries using equipment developed by Nutridan Engineering A/S of Denmark; collection in Denmark for 1984 was estimated at 10,000 metric tons (Widriksen 1984).

The blood is collected under sanitary conditions and then cooled after which it may be processed as whole blood or fractionated into plasma and red blood cells by centrifuging. The fractions are preserved by freezing or spray drying. In Denmark, frozen plasma (protein content $\sim 8\%$) sells for $0.18 per lb. and frozen hemoglobin sells for $0.045 per lb. The plasma is used mainly in the meat industry plus in the baking and pharmaceutical industries. Hemoglobin is added to foods as a coloring agent and is especially used as mink feeds (Widriksen 1984).

One company in the U.S. collects beef blood and processes it into plasma and red blood cells which are both spray dried. The dried plasma sells for $2 per lb. and is exported for use as a meat binder.

Color of blood is objectionable in many foods and the color is also unstable. Emulsifying blood with fat and protein (e.g., casein) decreases the intensity of the color; such emulsions are used in sausages. Although high in content of most of the essential amino acids, whole blood proteins are deficient in isoleucine and low in methionine and cystine. The chemical score for whole blood relative to whole egg proteins is only 14 because of the low isoleucine content. Availability of blood depends on an established meat industry, hence blood is not a likely protein source in countries where the diet is primarily vegetarian unless the product is imported.

Fish Protein Concentrates (FPC)

These products are made from fish or other aquatic animals by removal of water and, in some cases, of oil, bones and other constituents. Three types of FPC (types A, B and C) were defined by the Protein Advisory Group (1971). Type A FPC's are bland, colorless powders with minimum protein contents of 75% (on a 10% moisture basis) and a maximum lipid content of 0.5%. Type B FPC's have color and flavor, a minimum protein content of 60% and no-limit on lipid content. Type C products are essentially fish meals produced under hygenic conditions but have characteristic fishy odors and flavors that limit their use as food supplements.

Nutritive value of FPC is good; PER of type A FPC is equal to or better than casein (Sidwell et al. 1970). Blends of FPC with a variety of vegetable proteins gave better growth responses than the unsupplemented plant proteins.

Extensive studies on FPC's were conducted in the United States, Canada and elsewhere in the 1960's and 1970's as summarized by Finch (1977). In the United States, a major effort was made by the Bureau of Commercial Fisheries to develop a solvent extraction process to prepare type A FPC. Isopropyl alcohol was selected for use in a countercurrent extraction procedure to remove lipids and water from comminuted fish. Solvent was removed by vacuum or atmospheric drying followed by steam stripping to reduce residual solvent to below 0.025%. Unavailability of fish, residual odors and flavors, lack of markets for FPC, and poor functional properties (insolubility, gray color and gritty texture) were among problems encountered in the development work (Crisan 1978, Finch 1977). Several commercial operations were started but none are operating at present. In Japan, Niigata Engineering and the Fisheries Agency have developed a new FPC called Marinbeef (Hannigan 1983). Deboned fish is washed with sodium bicarbonate, kneaded into a paste which is then extruded into chilled ethanol to coagulate the proteins and remove lipids. The product is finally extracted with additional ethanol, desolventized and dried to yield greyish granules. The hydrated FPC is recommended as a meat extender but contains residual alcohol that is removed by washing the FPC during the hydration step prior to use.

Single-Cell Proteins

This term refers to the dried cells of microorganisms, such as algae, actinomycetes, bacteria, yeasts, molds and higher fungi produced in large-scale culture for use in foods and feeds (Lipinsky and Litchfield 1970, Litchfield 1977, 1978, 1983). Algae, bacteria, yeast and fungi are produced commercially in various countries around the world.

Algae. The algae are a diverse group of mainly aquatic plants ranging from microscopic single-celled organisms (micro-algae) to the giant seaweeds (macro-algae) found in the Pacific Ocean (Robinson and Toerien 1982). The macro-algae contain only 5-30% crude protein and are not consumed in large quantities; among these the brown algae are harvested for their trace elements and hydrocolloids (e.g., the alginates). The contribution of macro-algae to future protein supplies is considered to be only marginal.

Micro-algae are cultivated under conditions that range from very primitive (*Spirulina* growing on fermented cattle manure) to very sophisticated (*Chlorella* growing in ponds under controlled conditions). *Chlorella* species are grown in ponds or tanks utilizing sunlight as an energy source (photosynthesis) and carbon dioxide as the carbon source. Alternatively, *Chlorella* can also be grown in the dark if carbon and energy sources are provided (e.g., glucose) along with a nitrogen supply such as urea. Harvesting consists of initial concentration of cells, dewatering into a slurry and finally drying. *Chlorella* are presently grown in Taiwan and Japan and are converted into tablets that are sold as health foods.

Spirulina are grown in Lake Texcoco in Mexico utilizing the bicarbonate that occurs naturally in the alkaline waters of the lake. Consumption of *Spirulina* as a protein source dates back to the ancient Aztecs.

Protein content (corrected for nucleic acids and other non-protein constituents) for micro-algae ranges from 30 to 70%. The proteins contain a fairly well-balanced supply of essential amino acids except for methionine (Robinson and Toerien 1982). Toxicity studies with humans are limited in number; some workers reported gastrointestinal problems whereas others observed no adverse effects (Lipinsky and Litchfield 1970). Further studies are clearly needed to resolve questions of safety, especially at levels where algae would supply significant amounts of dietary protein. Accumulation of toxic minerals and polycyclic hydrocarbons are other potential problems if algae are grown in highly industrialized regions. Nucleic acids may pose a hazard because of urinary stone formation and gout.

Near-term potential of algae as a major protein source is not encouraging. *Chlorella* produced on a small scale in Taiwan and Japan at $5400-10,800/ton preclude use of such material except for a limited health food market (Robinson and Toerien 1982). By comparison, edible soy flour currently sells for only $240/ton. The green color and bitter flavor of some algae also severely limit their use. Algae production is practical only in areas below 35° latitude with sunlight most of the year. In addition, costs of harvesting and processing of algae into edible grades make them uncompetitive from economic and marketing standpoints (Litchfield 1977).

Yeasts. Baked products leavened by *Saccharomyces cerevisiae* (baker's yeast) have a long history of useage by man. More recently, other types of yeasts have been produced. For example, torula yeast *(Candida utilis)* is being produced by aseptic fermentation of grain alcohol in the presence of ammonia, water, air and minerals. The yeast is harvested by centrifuging, pasteurized and spray dried to yield a powder containing about 50% protein.

The edible grade product is used for its functional properties—flavor enhancement of meats, and binding of fat and water—as well as for protein fortification of foods (Anon 1977, Litchfield 1983).

Torula yeast is also produced from sulfite waste liquor and the yeast, *Kluyveromyces fragilis*, is prepared by fermenting cheese whey (Litchfield 1983). Baker's yeast (*S. cerevisiae*) is being produced by a continuous process from cottage cheese whey after lactose hydrolysis with immobilized lactase (Anon. 1984). Crude protein content of yeasts falls into the range of about 42-66% and essential amino acid compositions meet or exceed the 1965 FAO pattern except for methionine (Lipinsky and Litchfield 1970). Yeasts have found a niche in the protein ingredients market but primarily as flavoring ingredients. High selling prices (Table 5) place yeasts at a disadvantage as a protein source compared to conventional sources such as soybeans. Nucleic acid contents of yeasts (6-11%) further limit their utilization as protein ingredients (Litchfield 1980); removal of nucleic acids by processing would increase production costs.

Prospects for yeasts in the near future as primary sources of proteins are not very promising. Their uses are likely to be limited to protein supplementation and to functionality such as flavor enhancement (Litchfield 1983).

Soybean Proteins

Foremost among the plant proteins being used as food ingredients are those of soybeans. These proteins are incorporated into a large variety of foods and are no longer considered new or novel by the food industry. Three major forms of edible soybean proteins differing in protein content are available commercially (Smith and Circle 1972): (a) defatted flours and grits containing 50% protein; (b) concentrates with a protein content of 70%; and (c) isolates with minimum protein contents of 90%. Typical compositions for these products are given in Table 6. Dehulled, defatted soybean flakes produced under sanitary conditions are the starting material for the three protein forms. In addition to protein, defatted flours and grits contain water-soluble oligosaccharides, water-insoluble polysaccharides, ash, and a variety of minor components such as phytate, saponins and isoflavones. Protein concentrates are made by extracting defatted flours or flakes to remove the water-soluble oligosaccharides plus ash and some minor constituents. Isolates are processed one step farther to remove the oligosaccharides and the polysaccharides to yield essentially the storage proteins. Increased processing needed to prepare concentrates and isolates makes these products more expensive than flours and grits (Table 5).

Historically, numerous rat feeding studies have shown that soybean proteins have a good balance of essential amino acids except for methionine which is the first limiting amino acid. The relatively low contents of methionine and cystine likewise influence the chemical score of defatted soy flour which is 70 relative to whole egg protein (FAO/WHO 1965). Recent tests with adult

Table 6 Typical Compositions of Soybean Protein Products[a]

Constituent	Defatted flours and grits %	Protein Concentrates %	Isolates
Protein	56.0	72.0	96.0
Fat	1.0	1.0	0.1
Fiber	3.5	4.5	0.1
Ash	6.0	5.0	3.5
Soluble carbohydrates	14.0	2.5	0
Insoluble carbohydrates	19.5	15.0	0.3

[a] Analytical values on a moisture-free basis (Horan 1974).

humans, however, indicate that the protein qualities of soy protein concentrate (Istfan et al. 1983A) and soy protein isolate (Wayler et al. 1983) are high and comparable to those of high quality animal proteins, such as milk and beef. Digestibilities of protein concentrate and protein isolates by humans fall in the range of 91 to 96% and are comparable to digestibility values for milk (Istfan et al. 1983B, Scrimshaw et al. 1983). Contrary to results with rat feeding studies, supplementation of soy protein isolate with L-methionine showed no beneficial effects in young men when protein intake was adequate (Young et al. 1984). Nutritional studies with rats appear to underestimate the nutritional quality of soybean proteins for human adults.

Comparison of the essential amino acid contents of soybean proteins with the new FAO/WHO/UNU (1985) amino acid requirements in Table 3 shows that defatted meal and isolate A meet or exceed all of the amino acid requirements for children and adults. Isolate B meets all requirements except for children 2-5 years old, but it comes within 88% of meeting the sulfur amino acid requirement for this group. Table 4 shows that 2 g of soy isolate/70 kcal of infant formula is more than adequate to meet or exceed the amino acids provided by human milk at an equivalent caloric intake. The new FAO/WHO/UNU requirements are clearly in better agreement with results obtained from studies with humans than the chemical score discussed earlier. This agreement indicates that previous amino acid requirements were too high.

Raw soy flour inhibits growth of rats and causes hypertrophy of the pancreas but these effects are abolished by moist heat treatment of the raw flour and are attributed to trypsin inhibitors and possibly other heat-labile factors in soybeans (Liener 1981). Enlargement of the pancreas in response to raw soy flour varies with species. The rat pancreas enlarges on a raw soy flour diet whereas the pancreata of pigs and monkeys do not, but the response of all three species to properly heated soy flours and isolates is little different from the response of a casein diet (Struthers et al. 1983).

Soy flour may cause flatulence if the level of ingestion is sufficiently high; flatulence is believed to be caused by the oligosaccharides, raffinose and stachyose (Rackis 1981). Defatted soy flours contain 5-6% raffinose plus stachyose (Eldridge et al. 1979). Conversion of defatted flakes to concentrates or isolates removes these oligosaccharides and eliminates flatulence.

Raw defatted soy flakes also have characteristic grassy/beany and bitter flavors that must be reduced in intensity to be acceptable in foods. The usual treatment to reduce levels of the unobjectionable flavors is moist heat but this results in products with low protein solubility. Consequently, defatted flakes are converted to isolates to remove the flavor components and retain protein solubility.

The soybean protein industry in the United States is well established and the various protein forms are manufactured by eleven companies.

Walter J. Wolf

Table 7 U.S. Producers of Edible Soybean Protein Products

Company[a]	Flours[b]	Concentrates	Isolates	Textured Flours	Textured Concentrates	Textured Isolates
AP	•					
ADM	•	•	•	•		
Cargill	•			•		
CS	•	•		•	•	
GPC			•			
HP	•					
LG	•					
PMS				•		
RP			•			•
AES			•[c]			
WF					•[d]	•[e]

[a] AP = AG Processing, Inc.
ADM = Archer Daniels Midland, Co.
LG = Lauhoff Grain, Co.
AES = A. E. Staley Manufacturing, Co.
CS = Central Soya,
GP = Grain Processing Corp.,
HP = Honeymead Products,
PMS = PMS Foods, Inc.,
RP = Ralston Purina,
WF = Worthington Foods, Inc.

[b] Includes grits and flakes.

[c] Protease modified forms.

[d] Meat analogs containing soy protein concentrate plus isolate.

[e] Meat analogs containing spun soy protein isolate.

258

At present only about 2% of the soybean crop is converted in edible protein, hence there is a large potential supply available. Since edible proteins sell for a higher price than soybean meal used for feeds, the edible protein markets could compete successfully with feed markets for the raw materials (defatted flakes).

Wheat Gluten

Gluten is the rubbery residue that remains after washing a wheat flour dough to remove starch and soluble constituents. On careful drying, a light-cream to tan colored powder is obtained with a protein content of 70-75% which is the article of commerce (Fellers and Mecham 1974). The product also contains 5-10% lipids plus residual starch and moisture depending on the completeness of washing and drying. Wheat gluten forms a viscoelastic mass when hydrated even in the presence of excess water. In bread doughs, hydrated gluten stretches to form elastic films that entrap carbon dioxide during yeast fermentation thereby developing a cellular network that sets upon baking and thus giving bread its unique structure. Gluten is used primarily for its functional properties. Major use is in baked goods in the United States and Australia, two of the largest gluten producing countries. In continuously mixed process bread it improves dough handling, ensures high loaf volume and gives a softer crumb. Addition of gluten to hamburger and hot dog buns improves hinge strength. In Japan gluten is used extensively in processed meats and meat analogs to provide chewy texture (Sarkki 1979). Gluten contains almost 40% glutamic acid and is deficient in lysine as is common with most cereal proteins. The PER of gluten is low—0.7 to 1.0—because of the lysine deficiency, consequently it is not desirable for protein fortification unless it is blended with other proteins high in lysine (e.g., soy flour). Gluten is manufactured by four companies in the United States; it is also produced in Australia, Japan, Canada, South America and Europe.

Peanut Proteins

Edible grade defatted peanut flours and grits were available commercially until about 1986 and contained about 57% crude protein (Ayres et al. 1974). Partially defatted flours, however, are marketed at present. Protein concentrates and isolates can also be prepared from peanut flours but they are not produced commercially in the United States. Peanut proteins are low in lysine, threonine, isoleucine and leucine and PER values for peanuts fall in the range of 1.5 to 1.8 (Lusas 1979). In common with cottonseed, peanuts are prone to contamination with aflatoxin as a result of infection by the mold *Aspergillus flavus* if the peanuts are not handled and stored properly. The level of aflatoxins permitted in U.S. peanut products is 20 ppb and the level of aflatoxins in peanut products averages 2 ppb (Lusas 1979). Defatted peanut flours contain about 3% of galactosyl oligosaccharides; a typical analysis indicated 1.04% raffinose, 1.46% stachyose and 0.39% verbascose (Tharanathan et al. 1975). By comparison, raffinose plus stachyose in soy flours amount to 5-6% (Eldridge et al. 1979). Peanut flours and grits have not been used very extensively, in part, because of their high price.

Leaf Proteins

Much research and development has been carried out on recovering proteins from green leaves which are the world's largest source of protein (Fiorentini and Galoppini 1983; Humphries 1982, Pirie 1982, Telek and Graham 1983). Ribulose diphosphate carboxylase is a major protein in leaves and thus the enzyme is one of the world's most abundant proteins. Because leaves are high in fiber content, the proteins must be separated from the fiber and concentrated. The first step is to rupture the leaf cells and then squeeze out the juice containing the proteins. Next, the proteins are coagulated by rapid heating to 85°C using direct steam injection or by adding acids. The coagulated proteins are then separated by filtering or centrifuging, washed and finally dried to yield the whole leaf protein concentrate. The resulting product is green because of its chlorophyll content and also contains lipids. Chlorophyll and lipids can be removed by extracting the wet curd with polar organic solvents (e.g., propan-2-ol or butan-1-ol) prior to drying.

Alternatively, the leaf proteins can be separated into two major fractions: (a) those found in the chloroplasts where the pigments are also located and (b) cytoplasmic proteins found in the cell sap. These fractions can be separated by several techniques including differential heating. For example, rapid heating to about 60°C followed by rapid cooling to 40°C coagulates the chloroplast proteins which are removed by centrifuging to obtain a green coagulum. Reheating the liquid phase to 85°C then coagulates the cytoplasmic proteins which are recovered by centrifuging, washed, pressed and dried to yield a light colored protein product. Extensive work with this process has been conducted on alfalfa (see review by Kohler and Knuckles 1977). Pirie (1982), however, recommends using the unfractionated leaf protein in its green form, thereby eliminating loss of other nutrients such as β-carotene and making processing much simpler.

Amino acid analyses of leaf proteins indicate that they are a good source of lysine, methionine, threonine and tryptophan. Feeding trials with various animals and more limited studies with humans agree with the analytical data and indicate that properly prepared leaf proteins are a good source of protein but not as good as milk or eggs. Moreover, the leaf proteins do not appear to contain antinutritional factors.

Leaf proteins are being produced commercially in the U.S. and several countries in Europe (Pirie 1982), but firm establishment of an edible grade leaf protein industry has yet to occur. For preparation of edible protein, processing is about as complex as that used for producing protein isolates from defatted soybean flakes, but only 3% of the original leaf solids are recovered as edible protein. Drying the final product must be conducted carefully to avoid blackening and hardening. Preparation of leaf proteins in temperate zones suffers from a lack of raw material except for the summer months; the processing plants thus are idle an appreciable part of the year. Whole leaf proteins isolated by heating have poor functional properties. They are insoluble, lack the ability to gel, are green and have a grassy flavor.

Polyunsaturated lipids in leaf protein are prone to oxidation and formation of phosphorbides from chlorophyll is a potential problem unless processing is carefully controlled. Phosphorbides cause photosensitivity in pigs.

Cottonseed Proteins

A defatted cottonseed flour prepared by screw pressing was available in the United States for over 30 years but production was discontinued in 1976. The product was used in baked goods and confections to provide functional properties such as dough machinability, controlled spread, reduced fat absorption and improved browning (Spadaro and Gardner 1979). Cottonseed proteins are low in lysine, threonine, isoleucine and leucine, thereby making them unsuitable as a sole source of dietary protein. Cottonseed contains gossypol, a yellow polyphenolic compound, that is toxic to monogastric animals but toxicity in humans has not been reported. Glandless cottonseed, lacking the pigment glands containing gossypol, is produced on only a limited scale. Gossypol and other pigments impart a rich yellow color to products such as doughnuts but create undesirable colors in other foods. Defatted flour made from glandness cottonseed became available in the United States in 1986 but large-scale markets have not developed yet. Additional antinutritional factors found in cottonseed meal include the cyclopropenoid fatty acids, malvalic and sterculic acids. In rainbow trout, these fatty acids cause cancer of the liver; similar effects have, however, not been demonstrated to occur in humans (Hendricks et al. 1980). Cottonseed intended for human consumption must be stored carefully because the seed is prone to infection by *Aspergillus flavus* and contamination by aflatoxins.

Sunflower Seed Proteins

Sunflower nutmeats have been eaten as a snack item for many years and roasted nutmeats are presently used in candies, cookies, and snack items, and as salad garnishes. Flours, concentrates and isolates have been prepared experimentally but they are not produced on a commercial scale (Tranchino et al. 1983). Edible flours are difficult to prepare because efficient methods are lacking for thorough dehulling of the seed; residual hulls are high in fiber content and also contain pigments. Defatted flours contain polyphenolic compounds including chlorogenic acid. The polyphenolic materials are prone to oxidation to form undesirable pigments. For example, isolates prepared from toasted sunflower meals may be brown to dark green depending on the processing history of the meals. The polyphenolic compounds can be extracted from defatted meals with aqueous ethanol but this increases processing costs.

Sunflower proteins are low in lysine and leucine and borderline in threonine and isoleucine contents. The low lysine content of sunflower proteins makes these proteins unsuitable for blending with cereals which are also low in lysine. PER values of sunflower proteins are about 2.0 as compared to 2.5 for casein (Sosulski 1979).

Rapeseed Proteins

Rapeseed contains 40-45% oil and 25% protein of high nutritive value but the usual seed varieties contain 3-7% glucosinolates (Ohlson and Anjou 1979, Tranchino et al. 1983). Glucosinolates are biologically inactive but on hydrolysis by the naturally occurring enzyme, myrosinase, the glucosinolates yield nitriles and isothiocyanates which are toxic. Consequently, several processes have been developed to remove the glucosinolates from the seed and to prepare rapeseed protein concentrates. In a process developed by AB Karlshamns Oljefabriker in Sweden, the seeds are cracked and dehulled and then heated to inactivate myrosinase, thereby preventing hydrolysis of the glucosinolates. Heating also denatures and insolubilizes the major proteins. In the next step the heated seed meats are leached with water to remove the glucosinolates; the proteins do not dissolve because of the prior heat treatment. The leached meats are then dried and extracted with hexane to remove the oil. The defatted protein concentrate contains about 57% protein and   0.2% glucosinolates (Ohlson and Anjou 1979). The heat treatment step and extraction with water are analogous to a process formerly used in the U.S to make protein concentrates from defatted soy flour. Rapeseed protein isolates have been prepared experimentally and shown to be low in glucosinolates (Tranchino et al. 1983). In another approach to the problem of glucosinolates, Canadian plant breeders have bred the oilseed, canola, that is low in erucic acid and in glucosinolates (Daun 1984). Glucosinolate content must be lowered by an order of magnitude before rapeseed meal is suitable as a food ingredient (Diosday et al. 1984).

Rapeseed proteins contain more cystine and methionine than soybean proteins and rapeseed protein concentrate has a PER of 3.0-3.5 as compared to casein with a PER of 2.5 (Ohlson and Anjou 1979).

Although extensive developmental work has been done, food grade rapeseed protein products are still not available on the market (Tranchino et al. 1983).

Winged Bean Proteins

The legume, *Psophocarpus tetragonolobus* or winged bean, grows almost exclusively in the humid tropics of Papua New Guinea and Southeast Asia. Prior to 1975 the winged bean received scant notice but a report issued by the National Academy of Sciences (1981) called attention to this edible legume and encouraged research on it. Consequently, a number of workers have reported studies on the winged bean in the past decade; Sri Kantha and Erdman (1984) and Kadam and Salunkhe (1984) have recently reviewed these new findings.

Various parts of the plant—tuberous roots, leaves, flowers, shoots and seeds—are reported to be edible. Compositional data for 240 accessions from 16 countries indicated oil and protein ranges of 7-22% and 21-46%, respectively, for the seeds; ranges of mean values for oil and proteins for accessions from six countries, however, were narrower being 16-20% and 33-38%, respectively

(Hildebrand et al. 1981). These compositions approximate those of soybeans. Like soybean proteins, winged bean proteins are high in lysine but limiting in the sulfur amino acids, cystine and methionine. Other similarities to soybeans include the presence of trypsin inhibitors (Yamamoto et al. 1983), chymotrypsin inhibitors (Tan et al. 1984) and lectins (Pueppke 1979). As in other legumes, these proteins are inactivated by moist heat (de Lumen and Salamat 1980; Tan et al. 1983, 1984). Winged beans contain tannins that are concentrated in the hull although significant amounts also occur in the cotyledons. The tannins inhibit trypsin and are stable to moist heat (de Lumen and Salamat 1980). Raw winged beans are toxic to rats; toxicity is attributed to the lectin (Higuchi et al. 1983) and is abolished by autoclaving, although growth inhibition still occurs after heat treatment of the bean (Chan and de Lumen 1982). A cooked winged bean-maize diet was used successfully to treat children suffering from kwashiorkor (Cerny and Addy 1973) but more extensive nutritional studies with humans are desirable.

In contrast to soybean proteins that contain 2S, 7S, 11S and 15S fractions, one study indicates that winged bean proteins consist of only two major fractions of 2S and 6S plus a small amount of 11S-like fraction with the 2S fraction predominating (Gillespie and Blagrove 1978). Another study reported only a 2.5S and a 6.5S fraction (Yanagi 1983). Winged bean seed globulins have an isoelectric point of about pH 4.5 and isolates can be prepared by adjusting aqueous extracts of winged bean flours to this pH (Dench 1982). Aqueous extracts of full-fat winged bean flour upon dialysis and freeze drying yield protein concentrates that contain 71.5% protein and 9% fat (Sathe et al. 1982).

Winged bean is considered a promising source of protein and oil for the tropics and yields about the same amount of these constituents per hectare as soybeans (Claydon 1983). Nonetheless, there are a number of problems to solve before the full potential of this legume can be realized. A major obstacle to overcome is the lack of large scale production of the seed. At present, winged bean is primarily a backyard garden crop and the pods are often consumed as a green vegetable. Present cultivars are climbing herbaceous perennials grown on poles, trellises or fences, but cultivars that lend themselves to mechanized cultivation would be desirable. Because of lack of large supplies of seeds, there is no industry established to process winged beans into oil and meal. Likewise, there is no established industry to process defatted meals and flours into protein concentrates and isolates. Laboratory scale studies indicate that isolate yields are low and that tannins or other pigments cause the isolates to be brown (Dench 1982). Overall, near-term prospects for winged beans as a significant source of refined oil and protein isolates are uncertain. A more likely use of winged bean is a more direct consumption in traditional forms as a vegetable and in products such as tofu (Sri Kantha et al. 1983) and miso (Saio et al. 1984). A major research effort would be needed to make winged bean a commercial crop (Kadam and Salunkhe 1984).

Coconut Proteins

Coconut protein products are not available commercially, but developmental work has resulted in two classes of products, namely, those high in crude fiber and those low in fiber (Hagenmaier 1979). The high fiber product or coconut flour is the residue remaining after coconut meat has been dried and the oil removed by extraction. Coconut flours may be low or high in fat depending on the degree of oil removal:

	Low fat	High fat
Moisture, %	5	5
Oil, %	0.5	15
Protein (N X 6.25), %	24	20
Crude fiber, %	10	8
Ash, %	5	4

Low fiber products are made from coconut milk. One such product described by Hagenmaier (1979) is Cocopro. It is made by centrifuging coconut milk to remove some of the oil and then spray drying the low-fat coconut milk. The resulting white powder contains 32% protein and 8% oil.

Coconut flour has a good nutritional value with a PER of 2.8 and does not contain any known antinutritional factors. Cocopro has a lower PER value of 1.8. Both coconut flour and Cocopro have been incorporated successfully into bread and other baked goods. The future for coconut proteins is uncertain because there is no commercial production, product specifications have not been developed, and regulatory agencies have not examined the products. Market research and further development work are needed to determine whether coconut protein products can compete successfully in the market-place with other protein sources such as soybeans.

Relative Economics of Animal Protein Production Versus Plant Protein Food Ingredients

Relative costs of animal and plant proteins have been compared from a number of viewpoints. For example, Bradley (1962) used the water required to produce plant and animal proteins as a basis of comparison. He estimated that it takes about 100 gallons of water per day to produce one pound of wheat, whereas to provide one pound of beef protein and fat requires 2300 gallons of water per day or a ratio of 1 to 23.

Christiansen (1948) calculated the number of days of protein requirements that will be provided for a moderately active man by one acre of land used to produce various plant and animal proteins. Selected examples from Christiansen's data are as follows:

Product	Days of Protein Requirements/Acre
Beef	77
Milk	236
Wheat Flour	527
Soybeans, edible	2,224

Clearly, soybeans and wheat are more efficient in providing protein than beef and dairy cattle. Soybeans provide about 29 times as much protein as beef, 9 times as much protein as milk and 4 times as much protein as wheat flour.

The relative economics of animal and plant proteins are also apparent from the data in Table 5. Milk and egg proteins sell in the range of $0.15 to $1.79 on an as-is-basis but on a protein basis these prices range from $1.07 to $2.60 per lb. By contrast, soy, peanut and wheat proteins range from $0.15 to $1.05 on an as-is-basis and on a protein basis the range is only $0.30 to $1.58 per lb. At $1.20 per lb. at 50% protein or $2.40 per lb. of protein, torula yeast is higher in price than the plant proteins. The price discrepancy between animal and plant proteins is large for the low-protein products. For example, nonfat dry milk containing 37% protein sells for $0.95 per lb. while soy flours containing 50% protein sell for only $0.15 per lb. The high protein products, acid casein and soy protein isolate, are similarly priced whereas sodium caseinate, which is often used in preference to the acid form, sells for $0.40 more per lb. Subsidies play an important part in the prices of milk proteins. In the U.S. dairy price supports are responsible for the high prices of nonfat dry milk as compared to a world market price of 38¢/lb and the absence of a domestic casein industry. Overseas, subsidies and low production costs help to keep casein prices as low as they are. European Community countries probably would not be major suppliers of casein to the U.S. if subsidies were not paid to their dairy industry (USDA 1981). The high price of peanut flour as compared to soy flour is also the result of support prices that make peanuts an expensive starting material.

Conclusions

In comparing the merits of the various proteins considered as alternates for the traditional animal protein ingredients, one can quickly narrow the field down to a relatively small number. Among the non-traditional animal proteins, blood proteins have limitations of color, amino acid composition (deficiency of isoleucine) and cost, if the blood is fractionated. Plasma, which has found some specialized uses (meat binders), sells at $2.20 per pound of protein in spray-dried form and at $2.25 per pound in frozen form. The other animal protein is FPC developed in Japan after unsuccessful attempts elsewhere, but full-scale development is still uncertain. The product is expensive, subject to fluctuations in supply depending on availability of fish delivery to the manufacturing site, contains residual ethanol from processing, and is insoluble, thereby having limited functional properties.

Single cell proteins have limited potential to be used as protein ingredients except as specialty items such as health foods (algae) and flavorings (torula yeast). In these specialty uses, single cell proteins command high prices; torula yeast products sold as flavoring agents sell for $2.40 per lb. of protein (Table 5). Use of torula yeast and other single cell products at high levels is restricted by their nucleic acid content. The limit for nucleic acid ingestion is 2 g/day.

Of the plant protein sources considered, soybeans are clearly the most promising alternative source of protein food ingredients. The industry is well established and a wide range of products is available in the form of flours, protein concentrates and protein isolates that are widely used in the food industry. In the U.S., for example, there are six companies that prepare flours and grits and four manufacturers of concentrates and isolates. Soy flours are limited to applications where protein solubility is not always required and where ingestion levels are sufficiently low to avoid problems of flatulence. Flatulence problems are minimized by using concentrates and isolates. Isolates are the most highly refined and most expensive of the three protein forms, but are widely available and have the greatest versatility. Isolates frequently offer the greatest savings because they can be added to foods at higher levels than flours and concentrates. The soluble and insoluble carbohydrates have been removed and isolates are available in over a dozen forms tailored for specific applications such as baked goods, infant formulas, confections, meat products and analogs, coffee whiteners, whipped toppings, milk replacers and beverages. Although originally developed in the U.S., isolates today are also manufactured in other countries—Japan, Brazil and Belgium.

The remaining plant proteins being used commercially—peanuts and wheat—are not major factors in the protein ingredient marketplace. Peanuts are too high priced to compete with soybeans and do not have a good balance of essential amino acids. Wheat gluten occupies a specialty market to supply desired baking characterisitics in bread items and a chewy texture in meat analogs.

Among the remaining plant protein sources that were considered here, leaf protein, cottonseed, sunflower, rapeseed, winged bean and coconut are still in the potential category, but commercial development in the near future is uncertain for several reasons. These uncertainties include questions of safety (gossypol and cyclic fatty acids in cottonseed and glucosinolates in rapeseed), quality (high fiber in sunflower seed and coconut, and color due to phenolic acids in sunflower seeds and chlorophyll in leaf protein) availability (winged bean) and economics (applies to all).

References

Anon. 1977. Continuous aseptic system produces torula yeast. Food Engr. 49(6), 95-97.

Anon. 1984. Immobilized enzyme and fermentation technologies combine to produce bakers' yeast from whey. Food Technol. 38(6), 26-27.

AOAC. 1984. Official Methods of Analysis, 14th ed., S. Williams, ed. Association of Official Analytical Chemists, Washington, D.C., pp. 877-878.

Ayres, J. L., Branscomb, L. L., and Rogers, G. M. 1974. Processing of edible peanut flours and grits. J. Am. Oil Chem. Soc. 51, 133-136.

Bayless, T. M., Paige, D. M., and Ferry, G. D. 1971. Lactose intolerance and milk drinking habits. Gasteroenterology 60, 605-608.

Bookwalter, G. N., Kwolek, W. F., Black, L. T., and Griffin, E. L., Jr. 1971. Corn meal/soy flour blends: Characteristics and food applications. J. Food Sci. 36, 1026-1032.

Bradley, C. C. 1962. Human water needs and water use in America. Science 138, 489-491.

Cavins, J. F., Kwolek, W. F., Inglett, G. E., and Cowan, J. C. 1972. Amino acid analysis of soybean meal: Interlaboratory study. J. Assoc. Off. Anal. Chem. 55, 686-691.

Cerny, K., and Addy, H. A. 1973. The winged bean (Psophocarpus palustris Desv.) in the treatment of kwashiorkor. Brit. J. Nutr. 29, 105-112.

Chan, J., and de Lumen, B. 0. 1982. Biological effects of isolated trypsin inhibitor from winged bean (Psophocarpus tetragonolobus) on rats. J. Agric. Food Chem. 30, 46-50.

Christiansen, R. P. 1948. Efficient use of food resources in the United States. U.S. Dept. Agric., Washington, D.C., Tech. Bull. 963, 98 pp.

Claydon, A. 1983. Potential of winged bean pods and their products in Papua New Guinea. Qual. Plan. Plant Foods Hum. Nutr. 32, 167-177.

Clein, N. W. 1954. Cow's milk allergy in infants. Pediatr. Clin. North Am. 1, 949-962.

Collins-Williams, C. 1956. The incidence of milk allergy in pediatric practice. J. Pediatr. 48, 39-47.

Crisan, E. V. 1978. Fish protein concentration (FPC). In Encyclopedia of Food Science, M. S. Peterson and A. H. Johnson, eds. Avi Publishing Co., Inc., Westport, Connecticut, pp. 266-271.

Daun, J. K. 1984. Composition and use of canola seed, oil and meal. Cereal Foods World 29, 291-294, 296.

de Lumen, B. 0., and Salamat, L. A. 1980. Trypsin inhibitor activity in winged bean (Psophocarpus tetragonolobus) and the possible role of tannin. J. Agric. Food Chem. 28, 533-536.

Dench, J. E. 1982. Extraction of nitrogenous material from winged bean [Psophocarpus tetragonolobus (L.) DC] flour and the preparation and properties of protein isolates. J. Sci. Food Agric. 33, 173-184.

Diosday, L. L., Tzeng, Y.-M., and Rubin, L. J. 1984. Preparation of rapeseed concentrates and isolates using ultrafiltration. J. Food Sci. 49, 768-770, 776.

Eldridge, A. C., Black, L. T., and Wolf, W. J. 1979. Carbohydrate composition of soybean flours, protein concentrates and isolates. J. Agric. Food Chem. 27, 799-802.

FAO/WHO (Food and Agriculture Organization/World Health Organization). 1965. Protein requirements. Report of a joint FAO/WHO Expert Group. FAO Ntr. Mtgs./UNU Rept. Ser., No. c37. FAO, Rome, 71 pp.

FAO/WHO/UNU (Food and Agriculture Organization/World Health Organization/United Nations University). 1985. Energy and protein requirements. Report of a joint FAO/WHO/UNU Expert Consultation. WHO Tech. Rept. Ser. No. 724, WHO, Geneva, 206 pps.

Fellers, D. A., and Mecham, D. K. 1974. Gluten. In Encyclopedia of Food Technology, A. H. Johnson and M. S. Peterson, eds. Avi Publishing Co., Inc. Westport, Connecticut, pp. 481-483.

Finch, R. 1977. Whatever happened to fish protein concentrate? Food Technol. 31(5), 44, 46-47, 49, 52-53.

Fiorentini, R., and Galoppini, C. 1983. The proteins from leaves. Qual. Plant. Plant Foods Human Nutr. 32, 131-146.

Gillespie, J. M., and Blagrove, R. J. 1978. Isolation and composition of the seed globulins of winged bean, Phosphocarpus tetragonolobus (L.) DC. Aust. J. Plant Physiol. 5, 357-369.

Hagenmaier, R. 1979. Experimental coconut protein products. J. Am. Oil Chem. Soc. 56, 448-449.

Hannigan, K. 1983. Fish protein concentrate replaces meat. Food Engr. 55(3), 54.

Hendricks, J. D., Sinnhuber, R. 0., Loveland, P. M., Pawlowski, N. E., and Nixon, J. E. 1980. Hepatocarcinogenicity of glandless cottonseeds and cottonseed oil to rainbow trout (*Salmo gairdnerii*). Science 208, 309-311.

Higuchi, M., Suga, M., and Iwai, K. 1983. Participation of lectin in biological effects of raw winged bean seeds on rats. Agric. Biol. Chem. 47, 1879-1886.

Hildebrand, D. F., Chaven, C., Hymowitz, T., Bryan, H. H., and Duncan, A. A. 1981. Protein and oil content of winged bean seeds as measured by near-infrared light reflectance. Agron. J. 73, 623-624.

Horan, F. E. 1974. Soy protein products and their production. J. Am. Oil Chem. Soc. 51, 67A-73A.

Humphries, C. 1982. Towards leaf proteins as a human food. In Developments in Food Proteins-1, B. J. F. Hudson, ed. Applied Science Publishers, London, Chap. 8, pp. 263-288.

Istfan, N., Murray, E., Janghorbani, M., Evans, W. J., and Young, V. R. 1983A. The nutritional value of a soy protein concentrate (STAPRO-3200) for long-term protein nutritional maintenance in young men. J. Nutr. 113, 2524-2534.

Istfan, N., Murray, E., Janghorbani, M., and Young, V. R. 1983B. An evaluation of the nutritional value of a soy protein concentrate in young adult men using the short-term N-balance method. J. Nutr. 113, 2516-2523.

Kadam, S. S., and Salunkhe, D. K. 1984. Winged bean in human nutrition. CRC Crit. Rev. Food Sci. Nutr. 21, 1-40.

Kinsella, J. E. 1982. Relationships between structure and functional properties of food proteins. In Food Proteins, P. F. Fox and J. J. Condon, eds. Applied Science Publishers, London, Chap. 3, pp. 51-103.

Kohler, G. 0., and Knuckles, B. E. 1977. Edible protein from leaves. Food Technol. 31, (5) 191-195.

Liener, I. E. 1980. Toxic constituents of plant foodstuffs, 2nd Ed. Academic Press, New York, 502 pp.

Liener, I. E. 1981. Factors affecting the nutritional value of soya products. J. Am. Oil Chem Soc. 58, 406-415.

Lipinsky, E. S., and Litchfield, J. H. 1970. Algae, bacteria and yeasts as food or feed. CRC Crit. Rev. Food Technol. 1, 581-618.

Litchfield, J. H. 1977. Single-cell proteins. Food Technol. 31(5), 175-179.

Litchfield, J. H. 1978. Microbial cells on your menu. Chemtech 8, 218-223.

Litchfield, J. H. 1980. Foods, nonconventional. Encyl. Chem. Technol. 11, 184-207.

Litchfield, J. H. 1983. Single-cell proteins. Science 219, 740-746.

Lusas, E. W. 1979. Food uses of peanut protein. J. Am. Oil - Chem. Soc. 56, 425-430.

McGee, H. J., Long, S. R., and Briggs, W. R. 1984. Why whip egg whites in copper bowls? Nature 308, 667-668.

McLaughlan, J. M. 1979. Critique of methods for evaluation of protein quality. In Soy Protein and Human Nutrition, H. L. Wilcke, D. T. Hopkins, and D. H. Waggle, eds. Academic Press, New York, pp. 281-297.

Ministry of Agriculture, Forestry and Fisheries. 1982. Japanese Food Consumption Yearbook. Association of Agricultural Statistics, Tokyo.

National Academy of Sciences. 1974. Improvement of protein nutriture, Washington, D.C., 201 pp.

National Academy of Sciences. 1981. The winged bean: A high protein crop for the tropics, 2nd ed., Washington, D.C., 48 pp.

Ohlson, R., and Anjou, K. 1979. Rapeseed protein products. J. Am. Oil Chem. Soc. 56, 431-437.

Pirie, N. W. 1982. Food protein from leaves. In Food Proteins, P. F. Fox and J. J. Condon, eds. Applied Science Publishers, London, Chap. 10, pp. 191-210.

Protein Advisory Group. 1971. Revised PAG guidelines for fish protein concentrates for human consumption. PAG Guideline 9, United Nations, New York.

Pueppke, S. G. 1979. Purification and characterization of a lectin from seeds of the winged bean, *Psophocarpus tetragonolobus* (L.) DC. Biochim. Biophys. Acta 581, 63-70.

Rackis, J. J. 1981. Flatulence caused by soya and its control through processing. J. Am. Oil Chem. Soc. 58, 503-509.

Ranken, M. D. 1977. Food ingredients from animal blood. Chem. Ind. No. 12, 498-500.

Robinson, R. K., and Toerien, D. F. 1982. The algae—A future source of protein. In Developments in Food Proteins-1, B. J. F. Hudson, ed. Applied Science Publishers, London, Chap. 8, pp. 263-325.

Saio, K., Suzuki, H., Kobayashi, T., and Namikawa, M. 1984. Microstructural changes in winged bean and soybean during fermentation into miso. Food Microstruct. 3, 65-71.

Sarkki, M. L. 1979. Food uses of wheat gluten. J. Am. Oil Chem. Soc. 56, 443-446.

Sathe, S. K., Deshpande, S. S., and Salunkhe, D. K. 1982. Functional properties of winged bean [Psophocarpus tetragonolbus (L.) DC] proteins. J. Food Sci. 47, 503-509.

Scrimshaw, N. S., Wayler, A. H., Murray, E., Steinke, F. H., Rand, W. M., and Young, V. R. 1983. Nitrogen balance response in young men given one of two isolated soy proteins or milk proteins. J. Nutr. 113, 2492-2497.

Shurtleff, W., and Aoyagi, A. 1984. Soymilk Industry and Market, The Soyfoods Center, Lafayette, CA.

Sidwell, V. D., Stillings, B. R., and Knobl, G. M., Jr. 1970. The fish protein concentrate story. 10. U.S. Bureau of Commercial Fisheries FPC's: Nutritional quality and use in foods. Food Technol. 24, 876-878, 880, 882.

Smith, A. K., and Circle, S. J. 1972. Soybeans: Chemistry and Technology, Vol. 1, Proteins, AVI Publishing Co., Westport, CT.

Sosulski, F. 1979. Food uses of sunflower proteins. J. Am. Oil Chem. Soc. 56, 438-442.

Spadaro, J. J., and Gardner, H. K., Jr. 1979. Food uses for cottonseed protein. J. Am. Oil Soc. 56, 422-424.

Sri Kantha, S., and Erdman, J. W. Jr. 1984. The winged bean as an oil and protein source: A review. J. Am. Oil Chem. Soc. 61, 515-525.

Sri Kantha, S., Hettiarachchy, N. S., and Erdman, J. W. Jr. 1983. Laboratory scale production of winged bean curd. J. Food Sci. 48, 441-444, 447.

Steinke, F. H. 1979. Measuring protein quality of foods. In Soy Protein and Human Nutrition, H. L. Wilcke, D. T. Hopkins, and D. H. Waggle, eds. Academic Press, New York, pp. 307-312.

Struthers, B. J., MacDonald, J. R., Dahlgren, R. R., and Hopkins, D. T. 1983. Effects on the monkey, pig and pancreas of soy products with varying levels of trypsin inhibitor and comparison with the administration of cholecystokinin. J. Nutr. 113, 86-97.

Tan, N. H., Rahim, Z. H. A., Khor, H. T., and Wong, K. C. 1983. Winged bean (Psophocarpus tetragonolobus) tannin level, phytate content, and hemagglutinating activity. J. Agr. Food Chem. 31, 916-917.

Tan, N. H., Rahim, Z. H. A., Khor, H. T., and Wong, K. C. 1984. Chymotrypsin inhibitor activity in winged beans (Psophocarpus tetragonolobus). J. Agr. Food Chem. 32, 163-166.

Telek, L., and Graham, H. D. 1983. Leaf Protein.Concentrates. AVI Publishing Co., Westport, CT.

Terpstra, A. H. M., West, C. E., Fennis, J. T. C. M., Schouten, J. A., and van der Veen, E. A. 1984. Hypocholesterolemic effect of dietary soy protein versus casein in rhesus monkeys (*Macaca mulatta*). Am. J. Clin. Nutr. 39, 1-7.

Tharanathan, R. N., Wankhede, D. B., and Raghavendra Rao, M. R. R. 1975. Carbohydrate composition of groundnuts (*Arachis hypogea*). J. Sci. Food Agric. 26, 749-754.

Tranchino, L., Constantino, R., and Sodini, G. 1983. Food grade oilseed protein processing: sunflower and rapeseed. Qual. Plant. Plant Foods Hum. Nutr. 32, 305-334.

United States Department of Agriculture. 1971. Textured vegetable protein products (B-1) to be used in combination with meat for use in lunches and suppers served under child feeding programs. Food and Nutrition Service Notice 219, Washington, D.C.

United States Department of Agriculture. 1981. U.S. Casein and lactalbumin imports: An economic and policy perspective. Economics and Statistics Service Staff Report No. AGESS 810521, Washington, D.C.

United States Department of Agriculture. 1982. Agricultural Statistics. United States Government Printing Office, Washington, D.C.

Warriss, P. D., and Rhodes, D. N. 1977. Haemoglobin concentrations in beef. J. Sci. Food Agric. 28, 931-934.

Wayler, A., Queiroz, E., Scrimshaw, N. S., Steinke, F. H., Rand, W. M., and Young, V. R. 1983. Nitrogen balance studies in young men to assess the protein quality of an isolated soy protein in relation to meat proteins. J. Nutr. 113, 2485-2491.

Widriksen, F. 1984. Personal communication. Nutridan Engineering A/S, Herlev, Denmark, May 21.

Wismer-Pedersen, J. 1979. Utilization of animal blood in meat products. Food Technol. 33 (8), 76-80.

Yamamoto, M., Hara, S., and Ikenaka, T. 1983. Amino acid sequences of two trypsin inhibitors from winged bean seeds [Psophocarpus tetragonolobus (L.) DC]. J. Biochem. 94, 849-863.

Yanagi, S. 0. 1983. Properties of winged bean (Psophocarpus tetragonolobus) protein in comparison with soybean (Glycine max) and common bean (Phaseolous vulgaris) protein. Agric. Biol. Chem. 47, 2273-2280.

Yannai, S. 1980. Toxic factors induced by processing. In Toxic Constituents of Plant Foodstuffs, 2nd Ed., I. E. Liener, ed. Academic Press, New York, Chap. 12, pp. 371-427.

Young, V. R., Puig, M., Queiroz, E., Scrimshaw, N. S., and Rand, W. M. 1984. Evaluation of the protein quality of an isolated soy protein in young men: relative nitrogen requirements and effect of methionine supplementation. Am. J. Clin. Nutr. 39, 16-24.

Chapter 8

The Importance of Traditional Quality for Foods Containing Vegetable Protein Ingredients

*Stephen G. Sellers**
*John W. Bennett**
*William Cole**

*Department of Anthropology, Washington University, St. Louis, Missouri.

Abstract

Success of new technology in the food industry—whether new production techniques or new ingredients in processed foods—ultimately depends upon consumer acceptance of the final product.

Isolated soy protein represents such a new technology in the food industry, especially in Western diets. Used in various new and reformulated traditional food products, this ingredient has variously met with success and failure—adoption into consumers' regular diets, or rejection after trial.

How do we assess consumer reaction? How do people select their food? What factors determine whether or not consumers will accept a new food? What have food industry companies learned that will help them use new technology to produce foods that consumers will find appealing as well as beneficial?

Provision of adequate dietary quantities of protein has attracted major focus from modern food producers. Other chapters of this volume detail those trends, and the parallel increasing demand for protein in developing countries. This chapter demonstrates that isolated soy protein is technically superior to other protein food ingredients in fulfilling protein requirements and in functional performance, and is well accepted by consumers.

The authors discuss the biological, psychological and sociocultural bases of the process of human food selection—the information requirements, expectations, evaluation and decision making processes. They then identify the factors that influence acceptance or rejection of foods with vegetable protein ingredients. These include conceptual evaluation, sensory expectations, identification, convenience, flavor, texture, health benefits, digestibility, social significance (prestige), perceived value, place in the diet and price.

Using these principles, the authors review consumer reaction to a diverse sample of more than a dozen foods containing vegetable protein ingredients. The case review covers food in six categories: processed and reformulated meats, baked goods, beverages, special markets, analogs, and cereal blends and flours.

Foods reformulated with vegetable protein ingredients are most successful when they maintain traditional quality of foods already accepted in the diet. Rejection of foods with vegetable protein ingredients occurs most frequently because of discrepant sensory qualities.

Where these principles have been followed, consumer acceptance of cereal and grain products (breads) reformulated with vegetable protein has been high in general.

Those foods which are adopted into consumer diets, and which achieve commercial success, most frequently do so on the basis of price, health and special functions.

Lower price has attracted consumers to products in which vegetable protein ingredients have replaced more expensive ingredients. Most of these have been meat products. The price advantage alone is not sufficient, however; maintenance of sensory qualities and nutritional content of the traditional food is essential.

Special attributes are seen in analogs, such as simulated meat products and non-dairy cream substitutes.

Overall, producers have two strategies available when introducing new foods. The first is to offer distinctively new foods. This approach encourages consumers to be open-minded in their expectations regarding taste and other sensory attributes. It is difficult, however, to make such new foods sufficiently attractive conceptually so that consumers will seek them out. This strategy had a low success rate among the foods reviewed.

The second strategy is to replicate familiar foods. This approach facilitates consumer acceptance at the conceptual level because the new product is recognized as a traditional food. Again, however, it is critical that the reformulated food maintain the expected quality of the traditional product or it will be rejected. Isolated soy proteins' ability to replicate traditional quality has been key to the expansion of its use in many countries of the world in the last decade.

Introduction

Attempts to introduce untried foods* are often met with indifference or rejection. After all, it is said, people are notoriously conservative about what they eat. Travelers returning from foreign countries, for example, frequently recount their unhappy experiences with "exotic" foods, while conversely, immigrants to this country retain their eating habits years after they have assimilated the culture in other ways.

Consumer acceptance of untried foods, then, is problematic. Why are some foods accepted and others rejected? What factors influence the adoption of new foods? The introduction of foods with vegetable protein ingredients, especially soy derivatives, provides a number of instructive examples of both adoption and rejection. For example, hamburger extenders and meat analogs were widely rejected by consumers following their introduction in the late 1960's and early 1970's. A common opinion of that failure was expressed thus in a newspaper review citing an industry analyst: "'Changing consumer eating patterns is difficult to do in the U.S.' Realistically speaking, most experts say, it will be a long time before a food commonly associated with cattle feed can replace old favorites in the American diet." [Wall Street Journal, 10/26/77]. Similarly, advocates of programs for malnutrition relief frequently express their exasperation that "consumers often appear to be unwilling to accept advice and changes which others regard as self-evidently beneficial and based on incontrovertible scientific work" (Wilson, 1961: 134).

Contrary evidence is easily cited, however: "History proves that diet can undergo both rapid and major change with easy, even eager, community acceptance. Corn is not indigenous to East Africa, but it has become the major staple in large areas of that region in only a few generations. Sweet cola beverages are accepted, even coveted, from the Amazon jungles to the Russian Steppes despite the absence of local traditions regarding soft drinks. In the United States, yogurt, traditionally a food of Balkan and Eastern European herdsmen, is a nutritional craze and serves in various forms as a substitute for ice cream, mayonnaise, and sour cream" (Winikoff, 1978: 60). Or again, taking a longer view of midwestern U.S. diets: "Buffalo, venison, and horse, rather than today's beef, pork, and mutton, were at one time the major flesh foods of Europeans in western America" (Schmitt, 1952: 185).

Eating habits *do* change, even in the U.S., but not always in the desired direction. A more accurate assertion might be that diet preferences thwart manipulation, because many of the factors that influence diet change lie well outside the sphere of direct control.

The purpose of this review is to identify the factors that have influenced the acceptance/rejection of foods with vegetable protein ingredients. Using

---

Throughout this paper the term "untried" foods will refer generically to both those foods which are identifiably new to the consumer and those which are reformulations of traditional foods (and may not be initially perceived as new). See page 7 for further definitions of terms.

concepts from theories of human food selection, we interpret the outcomes of more than a dozen cases of new foods containing vegetable protein ingredients. The cases represent a spectrum of food categories so that, while the sample may not be statistically representative of the distribution of all foods with vegetable protein ingredients, it does encompass the variety of food forms. For theoretical principles of food selection we have drawn on research in biology, psychology, economics, and the other behavioral sciences. Our own disciplinary background is anthropological.

Much research is devoted to food from the perspectives of nutrition, technology, economics, and other disciplines which focus on food as a physical entity. Our concern here, however, is with human food selection and diet change, that is to say *behavior*, and so we look to behavioral principles in order to explain food choice. Having decided on the nature of the terrain, however, we immediately confront a dilemna regarding direction. Eating is an individual activity, but it takes place in a social context. Where do we start? Can social behavior be considered the aggregate of many individuals, thereby making the study of individual behavior a representative starting point; or is the individual such an incidental product of his social milieu that "social facts," as Durkheim called them, can only be pursued at a level transcending individual action? Homans puts it this way: "The central problem of the social sciences remains the one posed, in his own language and in his own era by Hobbes: How does the behavior of individuals create the characteristics of groups?" (Homans, 1967: 106). While we do not pretend to resolve the question here, we are obliged to seek a pragmatic solution to it in order to proceed. We have chosen the individual consumer as a starting point, and considered the society a part of the context in which food selection and diet change take place.

The focus of human food selection is the individual consumer, but individual choice is informed and influenced by the society at large. The acceptance or rejection of a new food, then, involves a combination of individual factors --taste, preference, etc. -- and sociocultural influences -- prestige attributes, nutritional knowledge, and so forth. We have formulated below a model of diet selection, defining those factors that seem pertinent to new foods with vegetable protein ingredients. The model begins with the consumer's initial perception of an untried food and ends with rejection or adoption into the usual diet. It entails two evaluation phases, termed conceptual and sensory, and alternation between phases may occur during evaluation. Food concepts are composed of multiple value dimensions and we show that foods that are perceived to enhance diet worth are more likely to be adopted. Research suggests that whether a food is initially perceived to be "new", that is, previously unexperienced, as opposed to familiar or "traditional", affects the way in which it will be evaluated by the consumer. This distinction helps to explain why sensory quality has been a primary criterion in some cases and price, nutrition, or other sociocultural influences have predominated in other instances for foods with vegetable protein ingredients.

Methods

The identification of foods with vegetable protein ingredients as a distinctive class of foods is amenable to the study of diet change for several reasons. For one thing, they are products of recent food technology; thus, the information is current and time depth is not too great. Second, these foods cover a variety of forms and several common food categories so that whatever can be inferred about consumer response will not be unique to a particular kind of food. Third, within their relatively brief U.S. history (significant commercial production of soybeans in the U.S. dates from the 1940's) several marketing "mistakes" have occurred so that something can be learned from a comparison of successes and failures. Fourth, they exemplify several attributes or functions associated with food including improved nutrition, enhanced sensory quality, and lower cost. Thus it is possible to examine which food features appeal to customers under different circumstances. Finally, major economic and nutritional benefits can result from foods with vegetable protein ingredients if they are acceptable to consumers thus lending practical merit to the enterprise.

This review of case material covers products in 6 categories: processed meats, cereal blended flours, beverages, baked goods, analogs of familiar foods, and products for special markets. We sampled a variety of food forms, of consumer markets, and of time periods. Data were collected in 1983-84 through interviews, unpublished technical reports, and published research findings. Among those interviewed were experts in the food industry, in government agencies, and in academic research institutions. Some of the information presented here has been made purposefully vague at the request of informational sources. Every attempt was made to corroborate specific data through multiple sources.

Terminology

Although the terms are sometimes confusing, particularly because they can have different meanings in the contexts of popular conversation, the food industry, and in the phraseology of government policy, we have tried to be consistent in our use of referants. We use "food" in reference to edibles in the form in which they are obtained or eaten by consumers. Thus hamburgers are a food and so is ground beef, but probably not salt. "Ingredients", like salt, are food elements but not usually consumed alone. "Traditional" or "familiar" foods are those that are already a part of the diet of a culture or individual, with "traditional" being somewhat more established in the diet. "Untried foods" are either recognizably unfamiliar to consumers or, sometimes, reformulations of traditional foods using different ingredients. Such reformulations might be imperceptible to the consumer, although this is unusual, but for our purposes will still be considered untried because of their innovative ingredient composition. Those foods which are recognizably unfamiliar we have referred to as "new" to indicate that consumers themselves are aware of the novelty. Thus new and reformulated foods are subsets of untried foods as shown schematically in Figure 1. All of the foods reviewed here were considered untried as defined above, whether they were perceptibly novel or not.

"Vegetable protein ingredients" are protein derivatives from vegetable (basically inanimate) sources. In all such products, protein refinement has been a major objective of the processing, even though they generally contain other components. The functions of derived protein ingredients can be several, but the criterion which distinguishes them from other plant derivatives, like corn starch, is the emphasis on protein (Martinez, 1979). Many of the vegetable protein ingredients are derived from soybean meal; they are soy grits, soy flour, soy concentrate, and isolated soy protein. Soy ingredients predominate in our review of cases, but we chose not to exclude foods with other vegetable protein ingredients in order to take advantage of the comparison. For example, most formulations of Incaparina do not include soy ingredients but instead use cottonseed to achieve a high protein content. Similarly, some protein-enriched *pastas* (though not reviewed here) have utilized "high" protein wheat flour.

In summary, we have reviewed cases of *untried foods with vegetable protein ingredients* and sought to explain their acceptance or rejection. We selected a sample of illustrative cases, based first on the criterion of covering the range of such foods, and second on the availability of information.

Terms for Innovative Foods

Theories of Food Selection

This paper has been informed principally by concepts derived from the social and behavioral sciences, in particular, anthropology, social psychology and sociology. The anthropological theme predominates, due to the disciplinary orientation of the writers. However, this does not mean that all propositions are derived from the professional literature in this field. "Anthropology" constitutes more of an orientating vector for our work; above all, it implies an emphasis on the cultural background of human behavior.

Appendix A contains a review of theories and explanations of food preferences derived from disciplines other than the social-behavioral sciences, principally experimental psychology and physiology. We summarize our conclusions from that review below, the reader is directed to the appendix for detailed exposition. The position we take is that while some of the factors identified in research experiments with animals and humans are important, in general they lie in the background of human food preferences, and do not provide the decisive reason for any particular choice or rejection. It is our position that these "decisive" reasons are to be located in human behavior in interaction with other people - a society - and with ideas and values - a culture.

In attempting to explain human diet choice, researchers have often distinguished major sets of factors. Shack, in defining "those areas of concern in the matter of food preference that would seem to be most cogent in a study of any consequence" (1975:209), distinguishes "sociological factors" and "psychological aspects" of food preference. Sociological factors include food procurement, food as an ethnic marker, and advertising; while psychological aspects are childhood eating habits, obesity, and symbolism of food sharing, among others. A similar typology of "factors which determine choice," (Yudkin, 1956) distinguishes 3 categories: availability, social and physiological. Availability, for Yudkin, includes geography, economics, and food technology; social factors include religion, social class, nutrition education, and advertising; physiological includes heredity, allergies, acceptability and nutritional need. A third typology underlying or implied by much food research distinguishes physiological, psychological, and sociocultural determinants of human nutritional behavior (de Garine, 1972).

Crosscutting these determinants of diet is a critical distinction between consumer food choice and food supply. Individual choice, whether adoption or rejection, assumes the availability of alternative foods. Research on choice seeks to explain, for example, why consumers select one food rather than another and it treats availability as largely exogenous to diet selection. In contrast, other research specifically addresses provisioning and availability; for example agricultural technology, environmental constraints, and marketing mechanisms. Research at this level seeks to explain not so much why consumers choose what things they do, but rather what determines the range of choice. Individual preference receives treatment only as an incidental influence on diet patterns.

Because we are dealing in this paper with the acceptance/rejection of foods with vegetable protein ingredients we have focused on consumer choice and not on supply. Why, for example, did consumers prefer breads with soy ingredients but rejected meat extenders in the early 1970's? Several explanations are possible: inborn taste preferences, habit, nutritional beliefs, etc. We have reviewed several theories of diet selection and propose a composite model which includes cultural values and sensory perception. In reviewing the research pertinent to theories of food selection we distinguish 5 explanatory models—innate, behavioral regulation, conditioned learning, imitative learning, and individual selection in sociocultural context. It is the 5th of these upon which we rely most heavily to interpret most of our case material. However, all of the models contribute something to our understanding of food selection.

The innate model suggests that taste preferences are inborn and so rejection or acceptance depends simply on whether flavor corresponds to innate preferences. Some food selection behavior in humans as well as other animals does seem to have an innate component—for example the attraction to sweet flavor and rejection of bitter—but the model leaves much unexplained. The behavioral regulation model posits an unconscious mechanism which enables animals to correct nutritional deficiencies. Under conditions of nutritional imbalance, the organism is thought to generate "specific hungers" which then alter preferences toward the increased ingestion of foods containing the lacking nutrients. Again, some human food choice may correspond to behavioral regulation; an example, it has been speculated, might be geophagia. However, counter-examples can also be cited: the persistence of nutritional deficiencies in human populations and, conversely, the selection of foods for reasons other than nutrient content. Learning theories of diet choice are predicated on behavioral conditioning. The idea is that humans, like experimental rats, try foods and remember the consequences: it made them feel good, for example, or made them sick. The body response is then recalled and associated with the food at the next eating occasion. In fact, the conditioning stimulus does not have to be metabolically related to the digestion of the food, as is evidenced by rats given electric shocks as well as by humans who experience an unhappy event while eating a certain food. While learned habits undoubtedly influence food choice, not all learning, especially among humans, occurs individually as is suggested by the conditioned learning model. Some food habits and preferences are be learned through direct imitation of the eating habits of others. Imitative feeding behavior has been observed in nonhuman primates and seems common in young children. Still, while social imitation may be influential, the acquisition of food preferences among adults is more complex than simple mimicry.

Sociocultural Model

Humans add a new dimension to the problem of food selection since social learning or culture, as well as purposeful behavior only tangentially related to nutrition, may modify, enhance, or even confound the efficacy of whatever food selection mechanism are potentially built-in. "Humans", it has been said, "eat food, not nutrients." The shared learning, communication, and conceptualization that comprise culture constitute a powerful mechanism to enable humans to deal with their environment. Unlike the physiological and psychological factors described above (and detailed in Appendix A), cultural mechanisms function at the level of the social aggregate. As humans have evolved and societies have developed, specific and direct food selection mechanisms have to a large degree been forsaken for one that is more powerful, more complex, and, on the whole, more adaptive. "By experience, translated into and transmitted by social custom, man has learned what foods to select to maintain reasonable if not perfect health." (Yudkin, 1956). But, like many other human activities, feeding behavior surpasses strict biological necessity and therefore much of the explanation for food choice must be pursued beyond biological need. Neither hamburgers nor caviar are chosen *primarily* for their nutritional worth, but because they are ascribed additional cultural values.

Our model of individual food selection emphasizes sociocultural influences, although idiosyncratic and physiological elements are not excluded. It is simply that we do not consider that individual or physiological factors can explain food preferences in any complete fashion, if one wishes to understand real-world patterns of choice and rejection of foods. It elaborates previous models in several ways. First, the process includes a social level, which provides an interactive milieu for the individuals information and thus informs the content of individual beliefs, knowledge, and concepts. This content is *cultural*. Therefore we view the individual evaluation of foods to take place not solely by sensory physiological mechanisms but also through conceptual processes. Humans judge foods not only on the plate but also as mental abstractions. (We *know* worms are disgusting to eat, even though we may never have tasted them.)

The model, as broadly defined above, provides a background for our understanding of the individual's behavior in the context of an actual situation in which he must, or wants to, make a decision about a particular food. Now, the question is at what point in the process do we intersect the individual's behavior? While he is still "interacting" with other people? After he has interacted and now has clear-cut concepts about what foods to eat, or what not to eat? That is, people do not come to the food preference situation as *tabula rasas*; they usually come with a storehouse of ideas, concepts, feelings, values which have accumulated as a result of their past experience - i.e. their interaction with other people, reading of printed material, and so on.

Since we cannot -- and do not need to -- reconstruct the entire history of the individual's social experience, we can take him as an individual, or as a typical representative of a group or population, and then proceed to characterize his selection behavior on the basis of whatever information we might have on cultural factors that he may or should possess. Some of the research to be cited or reviewed has focused on such situations: food preferences in individual and groups explained by social status, by exposure to medical literature, or by whatever factor was found to be (or deduced to be) effective.

Prior to ingestion, the consumer identifies a food by locating it in his mental library. A decision occurs at this point: either the food is readily recognized and judged traditional or it appears unfamiliar and requires further evaluation by reference to prior experience and food knowledge. Overall, then, the model incorporates sociocultural influences and individual cognition into the theory of food selection. Its core is composed of 3 phases: identification, sensory evaluation, and conceptual evaluation.

The Model

Since we are focussing on the acceptance or rejection of untried foods, we describe the application of the model to untried food selection, although it should be applicable to food selection in general. Initially, the (potential) consumer is exposed to an untried food and identifies it by reference to known foods and food information. That is, the consumer draws on his mental catalogue of foods and seeks to match the untried food with previous knowledge garnered from social and cultural sources. The match may be fairly specific -- "Oh, this is bacon." --or only an approximation -- " It looks like some kind of soft drink." Sensory cues serve to help identify the untried food. The visual appearance, the label on the box, or the verbal description provided by a friend are the kinds of first indicators which serve to position the untried food vis-a-vis stored knowledge.

The individual also contributes to and derives information from, his society's cultural pool. The food concepts a consumer has in his head are created and modified through learning. Some of that learning results from personal experience; that is, information from the trial and evaluation of food is stored for future use. Much of the learning, though, is social and thus concepts, values, and attitudes are influenced by the society at large. A scientific discovery about the health hazards of a certain food can alter its value to a consumer and so too can its prestige acquired through association with a social elite. Large-scale diet trends and patterns at the social level, influence individual selection by modifying conceptual values of foods.

The model of individual selection assumes, for heuristic purposes, that food knowledge and concepts are fixed at the moment of any particular trial. In this way the consumer's acceptance or rejection can be accounted for by his attitudes at the moment. We recognize, however, that concepts are susceptible to change, especially as occasioned by society as a whole, and so a food once rejected might re-emerge as acceptable given a change in values. In short, we have depicted the selection process at a point in time even though in actuality consumer selection is continuously in flux, subject to social influences.

In the model, food concepts are of different sizes, sometimes overlapping, and segmented. The size is meant to suggest that some food categories are broader, or more extensive than others. Overlap indicates that some concepts share elements and that one may subsume the other; for example, whole wheat bread is a conceptual subset of bread and may overlap the health food concept. The segmentation signifies the several dimensions of meaning and value that comprise a food concept. For example, french fried potatoes call to mind certain form (long and thin), flavor (salty), place in diet (fast food, accompanyment to hamburgers), nutritional worth (low, or "starch"), etc. Food information, we suggest, is mentally catalogued by images or concepts to which pertain certain values and significance. These mental images are learned, shared, and influenced in the experience of the individual in a social world. In addition, they form patterns which will be relatively uniform in particular social groups.

Having identified an untried food, perhaps only tentatively, the consumer begins to evaluate it. Two kinds of evaluation take place: one is sensory the other is conceptual. In both phases the food is judged relative to the initial image. Thus, if the consumer has identified the food as bacon, sensory evaluation is based on the known flavor, texture, and other organoleptic qualities of bacon. In the conceptual phase, the untried food is judged relative

284

to the salient value dimensions of the initial image. If the consumer considers it "better" than usual foods in the diet from the same category—for example, more nutritious or less expensive—it is likely to receive continued consideration. If, on the other hand, the available information on the untried food indicates no preferential value on any dimension, then it is likely to be rejected.

The process of trying and evaluating foods continues through time. Eventually the consumer may adopt it as a part of his usual diet or, at some point in the process, discontinue eating it. In fact, it might be said that so long as alternatives are available, foods in the diet are constantly being re-evaluated and are subject to change. Evaluation of a food generally involves several trials during which conscious decisions are made by the consumer about whether it is preferable than former foods and whether the sensory qualities are appropriate or appealing.

Individual Selection of New Foods in Social-Cultural Context

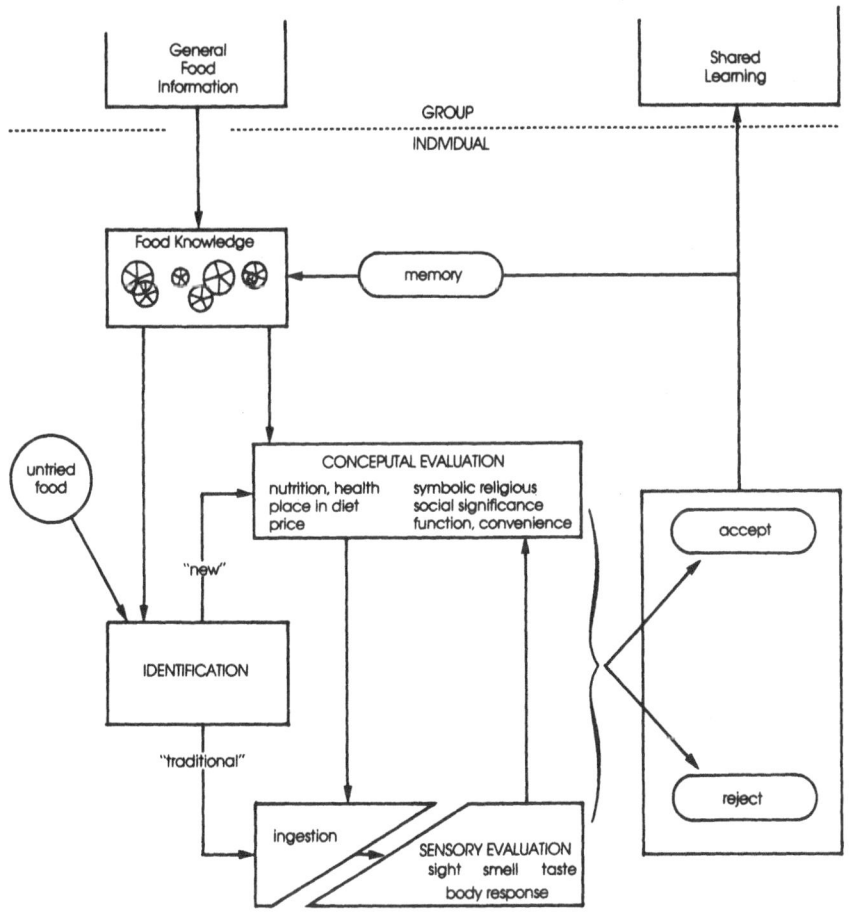

The Decision Flow

Those are the essential elements of the model, but there are several ramifications to be explained. The decision process is shown schematically in the figure on the next page. In following a specific food through the modified decision-tree, several points of divergence appear. At the level of initial identification, most untried foods will be identified by the consumer as "new"; which is to say that they do not exactly fit any preheld food concepts. The identification will be one of resemblance or association. More precisely, the individual usually does not make either-or choices but considers the position of the food on a kind of continuum of traditional to new (or conceptual congruence to absolute uniqueness). Colloquially, the range is from, "Sure, I know cabbage when I see it." to "What is that weird stuff lying next to the mashed potatoes?" Most untried foods will fall at neither extreme, but rather be considered relatively new or somewhat unfamiliar.

The distinction is important because the degree of proximity to preestablished concepts sets parameters for the proximate steps in evaluation. An untried food that is closely identified with a familiar food concept will be expected to closely resemble the food in its sensory qualities. Conversely, sensory evaluation may be less stringent for untried foods which are acknowledged to be unlike those experienced previously. On the other hand, food situated toward the novel end of the identification continuum will be subject to conceptual scrutiny: is it "better" in some respect than the usual foods in the consumer's diet? If not, why should he try it?

The distinction between modes of evaluation, and the notion that conceptual identification regulates routing through the selection process, are our own inventions. Together they enable us to accommodate findings about food behavior from different scientific disciplines that have so far been disparate. To depict in the diagram the difference in consideration, foods identified as traditional are shown to receive sensory evaluation first, while those identified as new are first subject to conceptual judgment. Characteristic behavior here may not be distinguished by temporal sequence so much as by critical emphasis.

Following a hypothetical case further through the model, recurrent trials allow the consumer to define more precisely the untried food and, eventually, to situate it concordantly in his usual diet or reject it. Some backtracking can occur during the process; for example, he might taste the food and decide, "I thought this was broccoli, but it tastes like shrimp, maybe it's some sort of seaweed." and so return to conceptual evaluation. Another elaboration in progress through the model is trade-off: the selection process entails some compromise or balancing of features. A texture which is slightly less desirable might be offset by substantially lower price; or a food that is somewhat inconvenient may be acceptable if it is compensated for by social esteem or religious worth. Trade-offs cannot be defined *a priori*, neither for individuals nor sectors of the society, but they can be inferred from consumer preferences.

Over repeated trials the untried food either becomes established in the usual diet or is rejected. Either way, the consumer has learned something about the food and that information is stored mentally to add to food conceptualization.

Decisions in the Evaluation Process

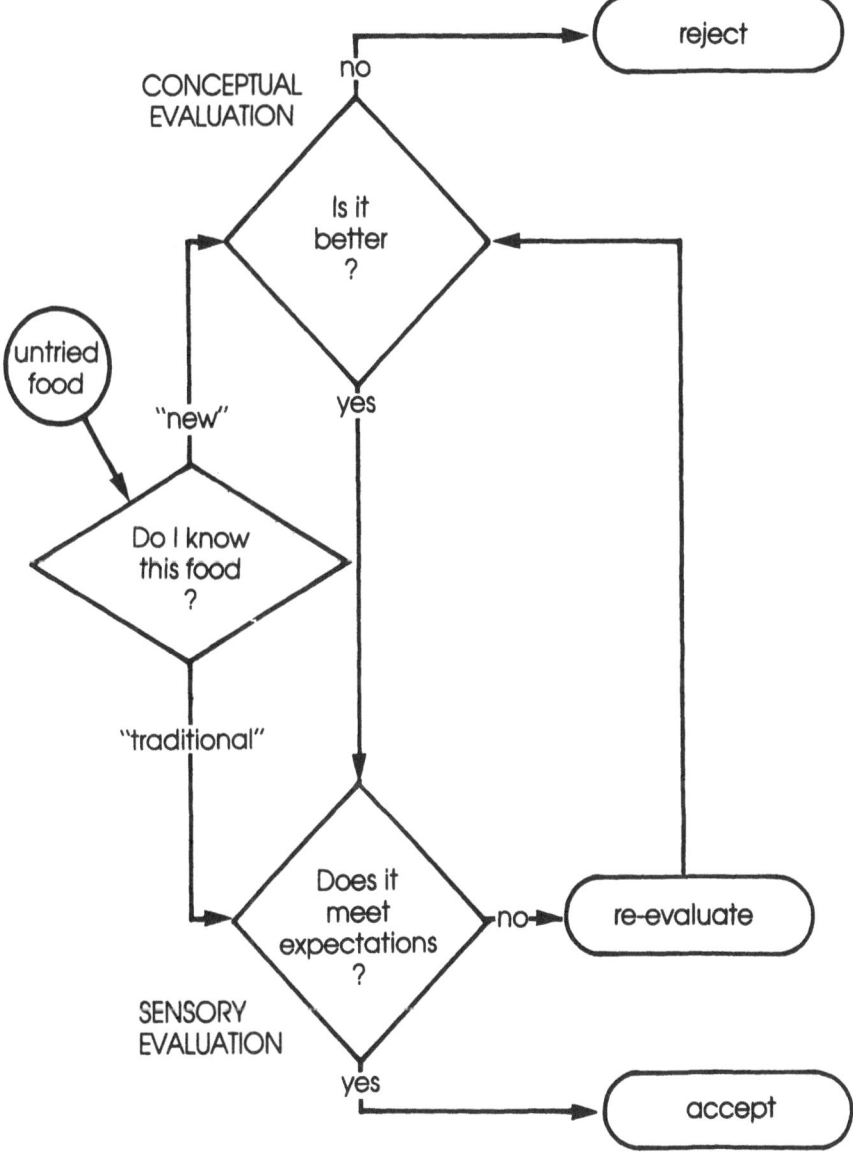

Identification

Conceptual evaluation presupposes a frame of reference and so the initial identification of a food can be critical for the ultimate decision. If a new food is being evaluated by a consumer, the outcome can depend very much on which foods it is compared to from the consumer's repertoire of knowledge. Whether the point of comparison is a high status luncheon food or a quick snack affects the decision. Untried foods that are perceived as new may be hard for the consumer to judge because the frame of comparison is uncertain. Promoters take advantage of uncertainty by attempting to "position" new products in order to skew the outcome favorably for their purposes, while consumers draw on their informational reserves to find a basis for comparison and evaluation. In contrast, foods that closely resemble familiar, traditional foods may bypass much conceptual evaluation because they are perceived to be equivalent (or identical) to an already accepted food.

Foods that are perceived as new or novel, are first situated by the consumer into a conceptual slot so that they can be evaluated. It serves no purpose to mentally compare ice cream to scrambled eggs if one will not substitute the other in the diet. The conceptual slot or image becomes more precisely defined as the evaluation proceeds and, by the time an untried food has been chosen for sensory evaluation some expectations have been formed about its composition and nature.

Sensory Evaluation

In sensory evaluation the consumer tries the food and decides whether to continue eating or not. In addition to taste, other senses—sight, feel, and smell—might also be involved so that sensory evaluation can entail several discrete steps and repetition of steps. Mouth evaluation is pre-eminent (Sharon, 1965; Fallon and Rozin, 1983) and taste is critical at this level. Moreover, we believe, perceived taste is judged relative to what the food *should taste like*. That is, an image has already been formed about the expected qualities of the food; if it does not meet expectations it is likely to be rejected.

Taste discrepancy, rather than unpalatability, then, is considered the major determinant of rejection at this stage. Even though the apparently innate palatability preference for sweet and salty, or aversion to bitter and irritants, may play some influential role, these responses have shown to be culturally reversible. And in any case, by the time it reaches his mouth a consumer will usually have already been aware of whether the food would be salty, etc. and have had the prior opportunity to reject it.

One of us had the opportunity to observe this phenomenon in his home:

> Because of a milk allergy, my son has always drunk a soy derived formula that looks like milk but tastes decidedly 'beany.' A friend was visiting us one day and, needing some milk for her coffee, borrowed a bit from my son's cup. At the first taste her eyes opened wide and she

nearly gagged. When she recovered, she asked how we could feed him such terrible tasting stuff! A week or so later, I was feeding my son at the table and mistakenly gave him the wrong cup of milk. He tasted the cow's milk, made a disgusted face, and said, 'I don't like this.'

The contrast illustrates, for one thing, that tastes are fundamentally acquired, but also how tastes that diverge from expectation are quickly rejected (Wilson, 1962). Figuratively speaking, they send off warning signals. Textural expectations operate in similar fashion. In food preference experiments during World War II soldiers were quick to reject "liquid meats" which had to be sucked through a vinyl tube (Peryam, 1963).

Conceptual Evaluation

Most human food decisions are made well before the food is on the table. The shopper selects foods mentally from a range of familiar choices before leaving the house, and prior to that, the range of alternatives has been acquired through lengthy experience out of a large variety of possible edibles. The ability to mentally conceptualize, evaluate and choose contributes to the efficiency of human food selection. Even more significantly, information is shared, communicated and accumulated in society. Thus knowledge about food -- what is poisonous, how to cook tripe, etc. -- is handed down across generations. Acquired food attitudes also define the range of choice. The human ability to decide and communicate about foods in the abstract is one of the great advantages that culture conveys for dealing with the environment. Humans are capable of making decisions about foods they have never seen before, or even of evoking nausea at the simple thought of eating certain things.

Conceptual evaluation is most easily understood in terms of rational choice (Cuthbertson, 1966): the potential food is compared conceptually to other known foods on the basis of preference of perceived value. Rational choice implies that consumers mentally compare alternative goods and select those that optimize satisfaction. In the simplest case, two foods might differ in only one respect and therefore the rational choice would be the one with greater preference value on that single variable. Because foods are endowed with multiple attributes, however, the choice is seldom so simple.

Culture endows objects and behavior with values and meaning. Foods, in particular, frequently carry a variety of values in addition to their nutritional worth. As concepts, foods are multivalent and their values derive from a variety of functions they serve in cultural systems. We recognize, for example, that French wine or soda pop and gefilte fish or catfish carry different values in different cultural settings. Foods can have nutritional value, religious worth, social status connotations, as well as price and sensory attributes. A full treatment of value dimensions appears in Appendix B.

Conceptual evaluation entails comparing an untried food to familiar foods on these multiple dimensions and, if it is to be selected, that it be judged significantly "better" than the familiar foods.

Summary: Valuation of Food Dimensions

The sociocultural model asserts that consumers will be more likely to adopt untried foods to which they ascribe higher value on one or more dimensions than foods in their existing diet. Six value dimensions seem especially pertinent to the adoption of foods with vegetable protein ingredients. In Appendix B we have reviewed information for each of these dimensions drawn from both Western and non-Western societies. The reader may wish to refer to that information as empirical substantiation for the following section in which we have formulated a brief proposition about the valuation of each value dimension insofar as cross-cultural generalization is possible.

- *social position*: Foods associated with superior status or with an emulated social group will be highly valued; foods identified with inferior status or rejected social group will be low-valued.

  Since foods do serve as markers of social boundaries, the positive or negative valuation of a food will depend on whether an individual seeks to identify himself with a social category, or, negatively, disassociate himself from a social category.

- *price*: All other things being equal, consumers in a market economy will prefer a food at a lower price to higher price.

  As personal income rises, preferred foods are incorporated into the diet while at the same time the proportion of the budget allocated to food diminishes. In Western societies, and other societies influenced by Western economy, meat has been highly esteemed and meat consumption rises with income. Grains, and other basic carbohydrate sources tend to respond less to income differences. Through the pricing mechanism, consumers trade-off not only foods relative to one another, but also against other possible expenditures.

  One effect of the expanding market economy in developing nations has been the replacement of subsistence production with store-bought foods in the diet. Consequently, malnutrition is increasingly a direct correlate of poverty.

- *nutritional beliefs*: Foods considered healthy or nutritional will tend to be positively valued relative to less healthy foods. It should be noted, however, that the nutritional dimension may be of low salience for a food (i.e., not a sought-for attribute in that class of foods) and that nutritional belief systems, even in contemporary U.S. culture, may vary considerably from scientific understanding of nutrition.

- *symbolic, religious*: Cross-cultural generalizations about symbolic valuation are constricted by the link, largely arbitrary, between form and meaning. Research on the food behavior of people in specific social settings and cultural traditions is required. However, two trends have been observed: food prohibitions are more common than prescriptions and most taboos concern animal products rather than plants.

- *function-convenience*: As a result of the premium placed on efficiency and time allocation in modern mass culture, foods requiring less preparation and consumption time may be positively valued. Under special circumstances ("gourmet dining"), however, the opposite prevails and elaborated consumption is preferred.

- *place in the diet*: New foods that fit a pre-existing niche in the diet tend to be positively valued over those that require restructuring of the diet pattern. Also, valuation is less rigid for foods that are peripheral in the diet (snacks, condiments) than for those that are central, so that the major meal, or principal plate, is more resistant to change.

Our theory suggests that new foods with superior valuation on one or more of these dimensions are most likely to be acceptable to consumers. Adoption is *not* guaranteed, however, because consumer evaluation *also* entails sensory trials and food expectations. Still, the prediction implied here is that those foods with vegetable protein ingredients that have had positive valuation on these dimensions should have a better acceptance record. We now turn to the review and interpretation of case material for new foods with vegetable protein ingredients.

Case Material

The following review of cases of foods with vegetable protein ingredients is divided into 6 categories: processed and reformulated meats, baked goods, beverages, special markets, analogs, and cereal blends and flours. Descriptive details of each case are summarized in the table beginning on the next page.

Table 1

	Time & Place	Form & Composition	Adoption Rate Use Level	Market Sector & Delivery
PROCESSED MEATS				
1.a. ground beef extender, Beef Patty Mix	Minneapolis 1973 - 1974.	ground beef patties with textured vegetable protein flour.	declined from 30% to 10% of market in 6 months, discontinued.	family supermarket purchasing
1.b. Jewel markets	Chicago 1979 to present	ground beef patties or chubs with isolated soy protein.	steady increase to over 25% market	family supermarket purchasing
2.a. luncheon meats	a latin American country 77 - 79, '81	hot dogs, bologna, etc. with soy flour, later isolated soy protein.	initial rejection, recent high acceptance	family supermarket purchasing
2.b. processed meats, pumped hams	Japan since early 1970's	formed hams	50% market	family consumption
3. fish paste products (kamaboko, chikuwa)	Japan 1977 to present	traditional fish paste items	20 to 30% market	family
BAKED GOODS				
1. Soja-Brot	Germany 1977 -present	breads and rolls, 10-25% soy grit.	6 million breads in Germany 1979.	family consumption, marketed through existing bakeries and bread plants.

2. golden elbow macaroni	Brazil, 1975	ready to cook pastas, soy flour.	high taste acceptability but low sales.	marketing outlets to low income families.
BEVERAGES				
1. Vitasoy	throughout Hong Kong, Japan, Indonesia, and Far East since 1940's.	soft drink in bottles or cartons, fruit flavored and dairy imitation. traditional beverage processed commercially since 1940's.	high: over 150 million bottles per year.	popular, healthy drink sold at markets and on the street.
2. Puma	Guyana 1969 - 73	bottled soft drink, artifically flavored.	used widely but briefly.	young adults, especially young men.
SPECIAL MARKETS				
1. diet foods, Nutri-Systems	diet centers in U.S., 1979 to present	ready to fix meals, low-calorie high-protein using isolate.	700 diet centers now in U.S.	overweight clients at weight-loss clinics
2. institutional foods	U.S. armed forces, 1979 to present.	extension of all ground beef in cafeterias with 20% soy concentrate.	high acceptability, total of 15 million lb. of ground beef per year.	all personnel at military base cafeterias.

Table 1
(continued)

ANALOGS				
1. breakfast meat substitutes	imitation bacon strips and sausages.	low acceptability.	U.S. 1972 to present	supermarkets, Seventh Day Adventists.
3. imitations dairy products, coffee creamers and whipped toppings.	nondairy cream substitutes, soy isolate from .8 to 2%	small, stable share of market.	U.S. since late 1960's (?).	supermarket and restaurant. Kosher dietary appeal.
CEREAL BLENDS, FLOURS				
1. Incaparina	flour to be mixed into a porridge by consumer	low volume, stable at about 200 tons per year.	Guatemala, 1961 to present; Colombia, 1963 to 1970's.	targetted malnourished women and children. Subsidized, price government regulated.
2. Cerex (corn soy milk)	flour to be mixed as a drink or integrated into other foods.	rates of acceptability over 85% in surveys.	Guyana, 1978 to present	targetted weaning infants, though consumed by all ages. Subsidized through food aid.

Processed And Reformulated Meats

Soy protein ingredients had been tried on a small scale in processed foods for a number of years, but it was not until 1973 that retailers of ground beef tried them on a significant commercial scale. The innovation was prompted by a sudden rise in beef prices in late 1972 and early 1973 which gave rise also to a shoppers' boycott of beef. The trial was brief; after an initial purchase few consumers returned for seconds. For a few weeks ground beef combined with textured soy flour comprised about 30% of the fresh hamburger market at grocery chains like Red Owl in Minneapolis where it sold for 20 to 25 cents less on the dollar than hamburger, but sales fell off quickly to about 10% of all ground beef sales.

In retrospect, the 1973 failure appears to have been due to a combination of factors, but foremost among them was dissatisfaction with the taste and texture of the extended ground beef. In addition to unappealing taste, the product discolored quickly, had a short shelf life, and produced flatulence. Finally, when beef prices fell in September 1973, the soy price advantage lost impact. What is most remarkable, perhaps, is that so many consumers were quick to try the extended ground beef. It demonstrates that, at the conceptual level, a significant number of consumers perceived potential advantages in the product. Distrust of reformulated foods and the stigma of adulteration, sometimes cited as reasons for rejection, did not constitute major initial barriers in this instance.

Textured soy flour was the ingredient most frequently used in 1973, and it was the oils and carbohydrates in that ingredient that were culpable for most of the disagreeable taste and texture that consumers identified. The new product was perceptibly different from ground chuck and so consumers' sensory expectation were unfulfilled.

In 1979 Jewel Food Stores began to sell a product called "Value Added Blend" containing isolated soy protein and selling for about 20 percent less than ground beef. Sales rose quickly in the first months and gradually thereafter so that Value Added Blend now constitute over 25% of all ground beef and over 50% of the commodity grade ground beef sales. That trend has continued even though beef prices fell in recent years. Trial sales by Ralston Purina of packaged, extended ground beef chubs and patties in several metropolitan area supermarkets in 1981-1982 followed a similar pattern reaching 27 to 30% of ground beef sales within the year.

The principal reason that recent sales have been sustained well beyond the 1973 introductions seems to be the use of a less detectable vegetable protein ingredient. Isolated soy protein is bland in flavor, preserves texture, and avoids other disagreeable qualities associated with the earlier soy flour. Taste tests carried out in school lunch programs for Ralston Purina in which whole ground beef was compared to ground beef reformulated with isolated soy protein and with textured soy flour found that a 30% mixture of hydrated isolate was indistinguishable from whole ground beef, while the textured soy flour mixture differed significantly from both. Attitudinal factors may also have influenced the greater adoption after 1979: the increasingly common sentiment through the late 1970's that beef and animal fat were unhealthy while grains were healthy may have lent nutritional worth to the soy protein isolate/beef combination in addition to its price advantage.

Processed meats manufactured by a major Latin American meat processor followed a similar pattern with the early rejection by consumers of reformulated products with textured soy flour and a more recent acceptance of those with isolated soy protein. In 1978 a major meat processing company in one Latin American country incorporated textured soy flour into a variety of processed meats sold in urban supermarkets under several brand labels. The items included hot dogs, *chorizo*, and *mortadela*.

Sales of these products fell off quickly after the introduction of textured soy flour as an ingredient. Since there was no modification of the labelling or advertising of the products, it appears that consumers had themselves detected changes in the foods and considered them undesirable.

In mid-1981, the same company decided to try isolated soy protein as an ingredient replacing about 20% of the meat in its hot dogs. A trial introduction in a secondary city was successful and soon after the company extended the practice nationwide with pressed hams and *mortadela*. By September 1982, when importation of isolated soy protein was suspended along with other imported goods because of deficits in the national trade balance, the reformulated hot dogs with isolated soy protein produced by this company and others were estimated to comprise over 67% of national sales.

The experience indicated that consumers were sensitive to product changes and responded to them quickly. In this Latin American country cereal flours and other inexpensive ingredients are routinely incorporated into processed meats. Perhaps because consumers expect processed meats to contain a variety of unknown ingredients, they are tolerant, but wary of, detectable changes. The processed meats containing isolated soy protein, unlike the earlier products, fell within the range of acceptable variation for what consumers recognized as the traditional hot dog.

In Japan, a similar tolerance for varied ingredients in processed hams may contribute to the standard practice of "pumping" hams to augment the supply. Isolated soy protein is used by about half the ham manufacturers for processing hams in Japan and the practice continues to grow steadily. By including isolated soy protein in the brine solution, 100 lbs. of meat can be converted into about 140 lbs. of sectioned and formed ham thus lowering production costs. Casein, egg albumin, and starch have been used for the same purpose, but isolated soy protein is advantageous in binding ability, texture, and water retention. Maintenance of protein content is an additional attribute.

The meat industry is a post-World War II phenomenon in Japan and its relative newness may partly explain the acceptance of ingredients in processed meats that for Westerners would be unexpected or discrepant. Some ham processing incorporates horse, mutton, rabbit, or minced fish paste which appear among the label ingredients but cause no apparent consternation.

Several traditional Japanese foods are based on fish paste *(surimi)* as a major ingredient; they include *kamaboko* which is a steam-cooked square, *agekama* which is deep-fried, and *chikuwa* which is broiled. The impulse to reformulate *surimi*-based foods using isolated soy protein occurred in 1977 when *surimi* prices skyrocketed because of the 200-mile fishing limit declared by the United States encompassing the fishing grounds used by the Japanese in the northern Pacific. Although price speculation also may have contributed to the rise in price of *surimi*, and prices fell once again six months later, a number of processors found *surimi* too expensive and consequently isolated soy protein reformulations of *agekama* and *chikuwa* reached 20% to 30% of their markets. For 100 parts of fish paste, both products traditionally contain 8 parts of potato starch, 30 parts water, and small portions of salt, sugar, monosodium glutamate, flavoring, and (optionally) minced vegetables. The usual reformulation technique entailed reducing the *surimi* to 70 parts and adding 30 parts hydrated isolated soy protein.

Isolated soy protein has several qualities which lend it to substitute readily for fish paste: it is light-colored, bland, and gelatinous when hydrated so that the appearance, taste and texture of the traditional food are maintained. There appears to be a slight attitude among Japanese consumers that reformulation with soy ingredients is not entirely "the real thing" -- only a few inroads have been made into *kamaboko* sales which is the most delicate and prestigious of this class of foods -- yet the facile switch between soy protein isolate and *surimi* and *vice versa*, when prices shift is a strong indication of their substitutability in the minds of consumers.

Another food in this category first appeared in northern Japan in 1977 and is a variant of *agekama*. Several producers, like Marubun Incorporated, created a deep-fried cake in the form of *agekama* but with a much higher vegetable content and containing an emulsion of isolate. An approximate breakdown is 30% hydrated isolated soy protein, 30% *surimi*, and 40% minced vegetables including carrots, onions, and others. It is distributed through the usual fish paste marketing channels and sold alongside the traditional foods on supermarket shelves.

This new version based on the traditional concept now shares about 5 to 10% of the *agekama* market. In addition to its price advantage it has the image of a tasty, healthy food that mothers can give their children. Its texture is somewhat softer than the traditional food and the vegetables are thought to be especially healthful. Clearly, in the minds of Japanese consumers, it falls within the general category of *surimi* based foods, and they look for it to have a white center because that is the sign of quality in traditional *agekama*. Yet it is not expected to precisely replicate any of the traditional products. This is an apparent example of a successful innovation which, either through inspired planning or trial-and-error, anticipated a diet trend.

Conclusions for Meat Products

In summary, some meat products containing vegetable protein ingredients have been accepted by consumers and others rejected. Taste and other sensory qualities seem to have been the major reasons for rejection of reformulated beef patties in 1973-1974 and of luncheon meats in a Latin American country in 1977-1979. In both instances a significant number of consumers did try the reformulated products, in one case (Latin America) unwittingly, in the other because of the price advantage relative to whole ground beef. Among the factors that have contributed to adoption of other products are lower price and nutritional quality. Most importantly, the accepted products -- reformulated ground beef patties in 1979, pumped hams, and Japanese fish paste products -- convey these advantages while matching the consumer's perception and expectations of traditional sensory qualities.

Baked Goods

Soja-Brot is one of several brands of breads using soy ingredients and sold in Western Europe since the late 1970's. Soja-Brot used 10 to 25% soy grit in breads and rolls giving them an appearance and chewiness like the whole-grain, "natural" breads now common in the U.S. They are marketed through existing baked-goods channels, the pre-mixed dough sometimes being supplied directly to local bakeries.

Breads with soy ingredients have achieved a small, but relatively secure, share of the European market. Their appeal seems to derive from their image as generically "healthy" though not specifically high in protein content. The soy ingredient also helps retain moisture and the sense of freshness.

Golden Elbow was developed by General Foods Corporation, produced and marketed by Industrias Matarazzo S.A. in Brazil. The objective was to produce a low cost, high (24%) protein food for poor, urban consumers. Since macaroni was a familiar food in the diet, high acceptance was expected. One of the promoters involved wrote that "Based on wheat-soy and corn flours - under normal commodity price conditions - it should be cheaper in raw material cost than ordinary semolina macaroni. It can be made on conventional pasta processing machinery and it has more than *sevenfold* the protein value of ordinary macaroni." [Bogyo, personal communication].

Taste acceptability tests in schools and elsewhere were favorable but the manufacturer discontinued production early in the first year. The major reason for discontinuation seems to have been that the producer felt that the product would sell equally well with wheat instead of soy flour as an ingredient and that there was no commercial advantage to be gained from the higher protein content. That is, the producer believed that consumers would be unwilling to pay a premium for macaroni with enhanced vegetable protein.

Conclusions for Baked Goods

Baked goods are a likely form in which to place vegetable protein ingredients. Unlike vegetable protein ingredients in meat products, which connote adulteration to some consumers, soy products were felt to be most appropriate in baked goods in a European survey. Because the principal ingredients are flours or meals, little sensory discrepancy is caused by vegetable protein ingredients and therefore taste is not a major barrier to consumer acceptance.

The difficulty, from a commercial perspective, is to enhance a product's appeal through the inclusion of vegetable protein ingredients; otherwise, why should consumers purchase it? Nutrient content, especially protein, is not a major attribute sought in most baked goods. Recently, however, there has been an upsurge in consumer interest in "healthy" breads in the U.S. and Europe so that vegetable protein ingredients can lend a commercial advantage. Also, their ability to retain moisture enhances perceived "freshness" to the bread.

Beverages

Vita Soy is the largest of several bottled soy drinks sold in Hong Kong, Singapore, Malaysia, Thailand and other parts of Asia. "By 1974 Vita soy passed Coca Cola to become Hong Kong's best selling soft drink, with sales of 150 million bottles a year" (Shurtleff, 1980: 28-36).

Vita Soy and the other soy-based soft drinks are a processed version of the traditional Chinese beverage, soy milk. Soy milk was developed in China about the 2nd century B.C. and has been consumed as either a spicy hot breakfast soup or as a warm, sweet beverage. Traditionally it was not used as a weaning food or infant milk substitute, perhaps because of digestive difficulties and flatulence. In the 1930's an American physician, Harry Miller, developed a process to steam distill, pasteurize, and homogenize soy milk and used it to feed Shanghai infants who were subject to malnutrition. Miller's purpose was to create an inexpensive, nutritious milk substitute for children. His technique was subsequently adopted by K.S. Lo in Hong Kong who established the Hong Kong Soya Bean Products Co. in 1945, and then by other companies in major Asian cities. In 1980 about 100,000 containers of 200 ml each were made daily by Japan's four largest soy milk producers. It is consumed by adults as well as children.

Acceptance of soy beverages in Asia has been quite high. Success of the commercial product seems to be attributable to having made a more digestible, portable, and inexpensive version of an already familiar, traditional food. In short, convenience is the major selling point.

Puma was developed by Monsanto Company in 1965 for introduction in Guyana and was modelled on Asian soy beverages. "Monsanto's goal was to produce an economically good-tasting beverage that would supplement diets deficient in proteins and essential vitamins and minerals." (Armenta, 1970: 289). Although it differed from conventional sodas in that it was not carbonated, had a lower sugar level, and was opaque, Puma rose to become the second best-selling soft drink after Pepsi-Cola, within 14 months of its introduction in Guyana in 1969. This occurred despite that fact that bottles were almost half the size of Pepsi (6-½ oz. versus 10 oz.) while selling at the same price and that local taste tests indicated a preference for Pepsi. Following its initial success similar drinks were introduced in Surinam (Samson by Coca Cola), in El Salvador (Fresca Viva by Pillsbury), and in Brazil (Saci by Coca Cola).

Much of Puma's quick popularity was probably due to astute marketing strategy. It was sold through the regular soft drink channels and was directed especially to teenagers and young adults. Advertising sought to establish Puma's image as "a healthy soft drink" and it seems that it attained connotations of virile strength so that it became popular as a beverage mixer.

Government tariffs on the soy protein and other imported ingredients seriously hindered Puma's production in 1973. It seems likely, however, though this is speculation, that an advantageous market position would have been difficult to sustain in any event. Initial sales may have resulted from the novelty and successful image of the drink but considering its costs, its dissimilarity with usual soft drinks, and the fate of analogous drinks in other parts of Latin America, a decline in sales seems to have been a likely eventuality. The experience of similar drinks in Latin American countries bears out this conclusion.

Conclusions for Beverages

In terms of consumer acceptance, the comparison of Vita Soy and Puma is instructive. Vita Soy was an enhanced, commercial version of a familiar, traditional product. The traditional version was already accepted in the Chinese diet; the commercial version was portable, storable, and reduced digestive difficulties. Puma, on the other hand, was a relatively new form of beverage. It was unlike the soft drinks that were familiar to consumers, and more expensive.

Special Markets

Several diet foods have been developed in recent years formulated with low total caloric content but high protein content. Some are sold directly to consumers while others are offered through comprehensive weight-reduction programs. Nutri-systems is one chain of diet clinics that offers a menu of low-calorie prepared meals. Dieters pay a premium for these meals and isolated soy protein is a major ingredient.

This is certainly a specialized market, perhaps even an aberrant consequence of U.S. dietary patterns. It is one of the few contexts in which it can be said that consumers are willing to pay a premium for protein. The market is contingent on U.S. eating habits and also on government-sponsored research regarding safe dieting practices.

Institutional consumption has already begun to have a significant impact on vegetable protein consumption and may presage things to come. In 1976, Wolf wrote that "In the institutional sector, the School Lunch Program is now the biggest consumer; about 25 million lunches are served daily. The School Lunch Program used about 9 million pounds of textured protein during the 1971-72 school year; in 1972-73 the amount is believed to have doubled." [Wolf, 1976]. Aggregate figures are not currently available, but local public schools report continued reliance on soy-extended ground beef.

Since 1979 the U.S. armed forces have extended ground beef in all military cafeterias with 20% soy concentrate (Schnakenberg, 1983). The measure was motivated primarily by cost savings and was preceded by evaluations of nutritional impact and taste acceptability. Army researchers report high levels of satisfaction with the practice. The newer soy concentrates now used as an ingredient are relatively bland and are attributed with the greater preference over the earlier flours. It may be, too, that the range of meal choices available to army personnel is institutionally constrained and that the extended hamburger would be less preferred on the open market.

The impact of extended ground beef in institutional settings on future taste preferences in the U.S. could be considerable. Since the 1950's beef has been the major meat in U.S. diets and since the mid-1970's more than 40% of beef has been consumed as ground beef. Because a majority of young Americans pass through the public schools and many through military service, receptivity to soy-extended beef in hamburgers, spaghetti sauce, and other foods is likely to be high in the future if only because of palate familiarity.

Conclusions for Special Markets

The influence of institutional constraints on individual food choice is exemplified by the cases of weight-reduction foods and U.S. Army hamburger. They underscore the point that foods are not chosen simply for palatability or preference, but that the decision about what to eat may be directed by the social context. The menu goes along with participation in the Army or weight-loss clinic. Whether the same selections would be made outside those social contexts is not certain, but it seems predictable that institutional foods are an advantageous outlet for vegetable protein ingredients. Primarily, the institution can reduce its food budget while maintaining the nutritional quality of its food.

Analogs

Meat analogs based on vegetable protein ingredients began to be introduced experimentally to U.S. consumers in the 1950's. Marketing efforts increased into the mid-1970's but have declined since then. Some examples of types of meat analogs are imitation hamburger patties, sausages, bacon, chicken patties, fish fillets, and scallops (Duda, 1974). Chunks or bits of meat analogs are also used in processed foods such as gravies, soups, casseroles, and pizza.

Since their introduction the meat analogs have passed through several marketing phases: the initial introduction to a specialized market, an attempt by larger producers to broaden the market, and, most recently, a return to a specialized market. The initial incentive for meat analogs has been vegetarianism. The two major producers in the 1960's -- Worthington Foods (now also Morningstar Farms) and Loma Linda --are closely identified with Seventh Day Adventists. Worthington Foods began to manufacture soybean-based meat substitutes as early as 1939 for Seventh Day Adventists (Ford, 1978). Their "meatless meats" fit the special needs of religious adherents who had grown up accustomed to meat in their diets. In theory it might be thought to ease the conversion to vegetarianism by not requiring a radical change in eating habits; even more, as suggested above, the analogs may serve as symbolic testimony to the adherent's vegetarianism.

A second marketing phase began in the early 1970's when several major companies became involved in producing meat analogs for a broader, commercial market. This period was also characterized by technological innovations which promised to make the analogs more palatable. Miles Laboratories purchased Worthington Foods and the Morningstar Brand to market a variety of products utilizing the technique of spun vegetable protein. General Mills, a major processor of textured soy flour, began to market a bacon substitute and imitation bacon bits as did General Foods. The appeal to a broader, nonsectarian market centered primarily on nutrition. The low-fat, low-cholesterol content was contrasted to standard bacon and sausages. Other attributes, such as low shrinkage, convenience, and storability, were also advertised.

Consumer acceptance of most meat analogs was low and, in the late 1970's, major producers abandoned much of the market. Worthington was sold by Miles Laboratories after financial losses. The imitation bacon strips have been discontinued by the major producers although the bacon bit substitutes continue to be sold. Most recently the meat analog business seems to be retrenching after having retreated from the attempts to reach a broad market with a variety of products. The new position is similar to the initial one, that is, the market is specialized, but it now includes some consumers who are attentive to cholesterol intake in addition to vegetarians. As a smaller, independent company, Worthington seems to have established some economic stability with this specialized market and so the meat analogs might be considered a partial success (or a qualified failure).

The failure of meat analogs to appeal to a larger commercial sector can be attributed primarily to unpalatibility. In their recent evaluation of meat analogs, the Consumer Union reported that "our sensory-test panelists, none of whom is a vegetarian, said that, with few exceptions, they would rather go without than eat an analog." (Consumer Reports, 1980: 363). Also, most meat analogs are no less expensive than the meat products they substitute so they have no price advantage. Evidently for most consumers, the nonmeat, low cholesterol content does not outweigh taste considerations when price is held relatively constant.

Several lessons can be learned from the meat analogs. First, sensory characteristics are crucial to food acceptance. The idea that analogs might be more acceptable to consumers because they resembled traditional foods (Welsh, 1979: 404) turns out to be incorrect. In fact, as suggested by our theory of food selection, tastes that are merely near to expectations, but not identical, tend to be quickly rejected by consumers.

Second, among the meat analogs, those that have had some general appeal, though limited, have been breakfast meats. This is consistent with the principle that peripheral meals are more amenable to change than those that are principal dietary events.

Third, meat analogs are viewed with some skepticism by U.S. consumers because of nutritional beliefs and attitudes. Although some shift away from the reliance on meat has occurred recently, prevailing health attitudes in the U.S. still emphasize meat protein as the primary nutrient source. Thus, in order to prefer meat substitutes, consumers must be persuaded that the nutritional quality is adequate. A trend in that direction seems to be taking place, yet meat analogs still confront nutritional skepticism on the part of most consumers.

Fourth, there does appear to be a market for meat analogs, but it is narrow and specialized. It consists of consumers who are willing to sacrifice taste preferences for a nonmeat product in the guise of a meat product. This, in itself, is somewhat enigmatic. Why should a consumer seek a vegetable food that pretends to be a meat instead of eating the same vegetable proteins in some other (and probably less expensive) form? We suggest that the symbolic significance to some vegetarians may account for this apparent enigma.

Dairy substitutes using vegetable protein ingredients include whipped toppings and nondairy coffee whiteners. Nondairy formulas for weaning infants also use vegetable protein ingredients and have been especially important for children who are allergic to cow's milk, however the acceptance problems for infants are different from adult foods and so we do not discuss them here.

Whipped toppings and coffee whitener are substitutes for dairy cream. The newer products often contain a low percentage (0.4 to 2.0%) of isolated soy protein in addition to palm oil, sucrose, and other ingredients. The vegetable protein is used as an ingredient primarily because of its function in forming and stabilizing an emulsion. In the whipped topping, the vegetable protein also functions to retain air.

The success of these cream substitutes can be attributed to several things. Among their advantages relative to cream, they are less expensive, can be stored for a longer time, can be easily transported, and contain no animal fat. In contrast to meat substitutes, the U.S. consumer has no expectation that cream will contribute significantly to nutrient intake and so there has been little question about its protein content. On the contrary, dairy cream is considered a nutritional liability because of its cholesterol content and therefore the nondairy creamer is preferable in that respect.

It seems likely that the cream substitutes have benefited from the gradual acceptability of margarine as a butter substitute. Margarine has gone through several phases of consumer acceptance, and, in 1957, surpassed butter in terms of *per capita* consumption in the U.S. (Nicholls, 197?). An early form of margarine was available in the early 1900's, but it was not widely consumed in the U.S. until World War II when it became a necessary substitute for butter. Accompanying its wider use there were several technological innovations which made it more palatible -- hydrogination, non-reversion, and precoloring. Until the Federal Margarine Act in 1950, margarine was still considered an inferior substitute for butter both popularly and officially, but in the late 1950's concern over cholesterol intake tipped the scales toward margarine and that trend has continued to the present.

The nondairy cream substitutes seem to have benefited from the consumers' gradual acceptance of margarine. They have not had to overcome doubts about their nutritional quality and have generally been viewed as having the same kinds of advantages relative to cream -- convenience, nutrition, and price -- that margarine did relative to butter. Moreover, they fit a well-established slot in the U.S. eating pattern. Most important, in our view, was that consumers had relaxed their taste expectations. Just as they had learned that margarine would not taste identical to butter, so too they were willing to tolerate some taste disparity between the nondairy substitute and dairy cream. Otherwise, we hypothesize, they would have met the same resistance that margarine did up to World War II.

Cereal Blends And Flours

Incaparina is the prototype of the high protein diet supplement (Shaw, 1972., 1975: 118). Subsequent products based on the same idea were Duryea and Colombiharina in Colombia, Cereal in Brazil, and Pro-Nutro in South Africa. Its principal designer, Nevin Scrimshaw, has remarked that Incaparina is intended to be more a concept than a specific formulation. The concept has been to manufacture a low-cost, high-protein beverage equivalent to milk [Scrimshaw, 1980] and several formulations have been employed to that end. Corn and cottonseed flour are usual major ingredients.

There is considerable discussion about whether, after 20 years, Incaparina should be considered a success or failure (Wise, 1980; Keegan, 1980). In Guatemala, where most distribution has occurred, it is hard to discern any trends in the nutritional status of the population that would justify the early hopes that Incaparina would alleviate malnutrition. The sales volume is miniscule in proportion to nutritional deficiences for much of the population [Wise, 1980]. On the other hand, the product has grown and continues to find a reliable market. Moreover, it might be argued that there should have been no expectation that Incaparina could resolve the larger social and economic issues underlying the high rates of malnutrition in Guatemala.

For our present interests, a more pertinent question is why has the demand for Incaparina not grown appreciably larger? One suggestion is that consumers did not like the taste or found Incaparina too dissimilar from the familiar *atoles* (gruels made from corn, plantain, or other foods) to be palatable. Another suggestion is that the monopoly franchise granted to the Guatemalan producer (Cerveceria Centroamericana) became a disincentive to effective marketing (Wise, 1980). The most persuasive argument is that Incaparina is too expensive for those who need it most; while holding little appeal to those who *can* afford it. In terms of protein-cost, Incaparina is advantageous relative to eggs or milk (Kracht, 1972), but compared to corn meal, which is the actual diet substitute as perceived by poor Guatemalans, Incaparina is considerably more expensive.

On the whole, Incaparina, like other cereal blends, is caught with an image problem that places it narrowly and uncertainly in the marketplace. Through advertising and health clinic distribution, Incaparina successfully acquired the image of a healthy food, like an *atol*, if not milk. Incaparina is widely perceived as a healthy food, especially for sick children. However, few people visualize Incaparina as a regular component of the diet - it is too expensive and, besides, *atoles* are not a daily diet item. At the same time it is not a prestige food (neither are *atoles* except on festive occasions) and so the populace that might afford Incaparina on a regular basis has little desire to buy it. The narrow sector of the market that Incaparina does fit is the urban, lower-middle class who, while not the neediest, are attentive to nutritional needs. In that narrow market range, however, Incaparina consumption is susceptible to fluctuations in the cost and availability of potential substitutes: cow's milk and charitable food donations. The critical growth constraint, then, seems to be not so much unpalatability or unfamiliarity but rather its position as a relatively expensive, low-prestige, but healthy, food situated between cheaper and more attractive alternatives. Competition from the malnutrition relief formulas offered free or at subsidized prices (as in Colombia), creates a nearly untenable position for products which seek to effect nutritional improvement while still achieving a return on investment.

Cerex is the brand name given to a corn soy milk formula marketed in Guyana since 1978 (Hopkins, Nichols, and Chin, 1983). Like other formulas developed in the late 1960's, it is intended for malnutrition relief and receives major cost subsidization, especially through the United States Department of Agriculture's PL-480 program. In Guyana, Cerex is sold in local food stores as a packaged flour and it can be mixed with water to make a drink or porridge. As for many food relief programs, infants, pregnant and lactating women are the targetted consumption sector.

Consumer acceptability of Cerex was evaluated in 1983 and so it serves as a recent, representative example of cereal blends. The major conclusion regarding Cerex is that it has very high consumer acceptability. Surveys of consumer attitudes showed that over 85% of families knew of Cerex and had tried it. Observations of purchase and consumption patterns revealed that many families regularly purchased Cerex and that it was consumed by the entire family. In terms of its original objectives, the product might be considered too successful in that the targetted infants and mothers had to compete with other consumers for the formula. Economic conditions in Guyana at that time were so disadvantageous that food availability in the markets was low and prices were high. Under those circumstances, with its subsidized price, Cerex became an attractive food for general consumption.

One inference to be drawn from the Cerex experience is that the taste and texture of the formula was not a drawback to consumption. In fact, few problems of rejection by consumers have been reported in the food assistance programs, though a sweetened version of corn soy milk has been developed to promote acceptability (Senti, 1974). On occasions of frank need, corn soy blend and wheat soy blend have been adopted quickly and the cost (or noncost) to consumers is usually quite favorable.

Conclusions for Cereal Blends

As far as the welfare agencies are concerned, a prime consideration for the need of a cereal blend must be its effectiveness in reducing or preventing malnutrition. For example, the amino acid fortification of wheat, rice, and corn, which seemed a promising approach in the early 1960's to the prevention of malnutrition, has been discontinued because little impact was discovered in field trials (Austin and Zeitlin: 1981). Similarly, welfare agencies must evaluate the effectiveness of cereal blends relative to other forms of food aid and other types of public health interventions such as innoculations for contagious disease or environmental hygiene.

From the consumer's perspective on acceptability, this review indicates that when in need, people will consume the blends if they are available. For the poorest, price becomes the major determinant of availability. It seems unlikely that the blends will become commercially viable without subsidization, except as infant formulas. The reason is not palatability, but rather preference: those who can afford a better diet that includes cereal blends do not see them as appealing foods compared to affordable alternatives. They connote low social status.

Cereal blend consumption in these situations is decided not so much by the consumer as by international agencies and national governments. The motives usually include both social welfare considerations and political favors since one or more governments will need to subsidize the product.

Conclusion To Data Review

This review of foods containing vegetable protein ingredients includes successes and failures across a variety of food categories. In isolating those factors which have influenced consumer acceptance or rejection we have distinguished those which we believe to have been especially instrumental and discount those which seem to have been less influential. In accordance with our model of food selection, we have highlighted consumers' identification of new foods, the pertinent criteria for conceptual evaluation, and sensory expectations. Having looked at factors on a case-by-case basis in this section, we derive some general conclusions in the next section.

Interpretation of Results

The theory of individual food selection in social-cultural context is based on an interaction between cognitive and physiological factors. Cognitively, the food should convey some perceived benefit to the consumer in order to be accepted. Physiologically, the sensory qualities of the food should satisfy expectations. The interplay between sensory familiarity and conceptual appeal frequently decides consumer acceptance or rejection. Moreover, we argue, the mental image or identification of an untried food charts the course by which evaluation occurs and so consumer perception of the food also influences selection. In the section below, we first summarize and categorize the case study results and then interpret them in terms of the model of food selection.

Summary findings for the foods reviewed are presented in the table on the next page. To facilitate the summary, product success was judged according to amounts consumed and to the rate of consumption over time -- whether rising or falling. Products that have either been discontinued, are falling off rapidly, or have been consumed much less than was intended, were considered to be failures.

Summary Of Findings

For the sake of simplicity our estimate of success/failure is dichotomized, with a code left for intermediate possibilities. However, real situations are rarely so straightforward. It is easy to judge as a failure a product withdrawn from the market, but one that continues to be consumed by a small sector of society, perhaps for reasons of religious dietary, might or might not be considered successful: low volume implies little acceptance but stability suggests regular demand. Such instances demonstrate that the intentions of the producer or supplier should play a part in the assessment. If the product was targetted toward a specific market sector, then the consumption rate *within that sector* would be the most appropriate basis for evaluation. Alternatively, a food may be acceptable to a wide range of consumers, but eaten only infrequently -- like cranberries in the U.S., for example. Again, the consumption rate should be judged in terms of producers' expectations.

For commercial products, producers and suppliers expect to make a profit. Some food products have failed not so much because of consumer demand or acceptability but rather because of production costs or other considerations on the supply side. Our evaluation of success/failure is not based on commercial profitability because several foods -- for example, cereal blends and U.S. Army extended ground beef -- do not compete in the commercial

Table 2

Acceptance-Rejection of Foods with Vegetable Protein Ingredients

Product	Success/ Failure	Untried/ Traditional	Why Accepted	Why Rejected
Meats				
Beef Patty Mix (1973)	-	u-t		sensory
Jewel Beef Patties	+	t	price	
Luncheon Meats (1977)	-	t		sensory
Luncheon Meats (1981)	+	t	price	
Pumped Hams	+	t	price, sensory	
Fish Paste	+	u-t		
Baked Goods				
Soja-Brot	+	u	health	
Golden Elbow	-	t		cost
Drinks				
Vita Soy	+	t	convenience	
Puma	-	u	health	cost
Special Markets				
Nutri-Systems Meals	?	u-t	health	health
U.S. Army Ground Beef	+	t	price, nutrition	
Analogs				
Breakfast "Meats"	?	u	health, religious	health
Non-Dairy Cream	+	u	health, price, convenience, place in diet	
Blends				
Incaparina	?	u	price, health	cost, status
Cereal	+	t	price	

Note: For Success/Failure, + = success, - = failure, ? = intermediate; for Untried/Traditional, u = untried, t = traditional, u-t = intermediate.

marketplace. Under other circumstances, these noncommercial foods might be judged nonsuccesses because they have entailed government intervention.

Summary of Results

Beef Patty Mix, like other extended ground beef products in the early 1970's, is judged a failure because it was withdrawn from the market soon after introduction. Consumers perceived it to be a substitute for whole ground beef and rejected it mostly because sensory qualities were too divergent from the familiar product, despite a lower price.

The Jewel hamburgers blend introduced in 1979 is judged a success because it constitutes a considerable and stable share of ground beef sales in this supermarket chain. The primary reason for consumer acceptance is lower price, relative to hamburger. Health considerations (lower animal fat) may be a secondary attribute. In contrast to earlier blends, sensory quality is satisfactory to consumers so the trade-off between price and taste is not disadvantageous.

The Latin American luncheon meats extended with textured soy flour in 1977 were quickly rejected by consumers and can be considered a failure. Evidently consumers had no idea that they were consuming a reformulated food until the unfamiliar taste alerted them. They believed they were consuming a traditional food. Thus, rejection was primarily on the basis of unfulfilled sensory expectations.

The same Latin American meats using isolated soy protein as an ingredient in 1981-1982 were acceptable to consumers and their price relative to competing luncheon meats led to renewed sales and are considered successful. The discrepant sensory qualities associated with the earlier product were absent when isolated soy protein was the ingredient.

Formed hams in Japan using isolated soy protein are judged a success because they constitute a considerable share of the market and are gradually increasing. Manufacturing costs are reduced and so price to the consumer is competitive while sensory qualities are favorable relative to formed hams that include other ingredients. Thus, in terms of consumer acceptance, price and sensory qualities are the attractive features.

Japanese fish paste products are judged a success because they constitute a significant and stable share of a very large market. Consumers identify these products with other *surimi*-based foods, but may not look for them to be identical to a particular traditional food. They have been commercially successful because of the lower price relative to traditional fish-paste products and, secondarily, because their inclusion of vegetable ingredients is seen as "healthy".

Soja-Brot has been successful in establishing a regular share of the European bread market. Its appeal to consumers is derived from the healthy image of "natural breads".

Golden Elbow macaroni failed as a commercial, high-protein replacement for wheat flour macaroni in Brazil. Its failure was due not to consumer rejection of sensory qualities, but because the manufacturer felt that production costs were high and that consumers would be unwilling to pay a premium for high protein macaroni.

Vita soy, like other soy-based soft drinks in the Far East, has been a remarkable commercial success. Such drinks are, in essence, commercial versions of a drink from the traditional Chinese diet and so they were already familiar to consumers. Their major appeal seems to have been convenience because they are purchased in bottles or cartons and are readily stored, transported, and available to the consumer. Price, digestibility and flavoring are other considerations.

Puma, like other soy-based soft drinks introduced into Latin American countries in the late 1960's, was eventually a commercial failure despite some initial success. Its appearance was unlike familiar soft drinks in Guyana. It appealed to conceptions of health and for a time consumers were even willing to pay a premium.

Nutri-systems foods can be judged only tentatively. They appeal to a small market of dieters who are willing to pay a premium price for low calorie/high protein foods. Their appeal is based on nutritional attitudes and beliefs and may be subject to informational changes in U.S. society.

Ground beef extended with textured soy concentrate in the U.S. army can only be considered a qualified success. The explicit attraction is reduced price. Even though the purchaser is not the direct consumer, taste tests conducted in the armed services indicate that sensory acceptability has met expectations, but only in the absence of alternatives. Transportability of vegetable protein has been a secondary consideration to the army.

Most analogs have had little commercial success, and those that have been successful have achieved it through specific attributes. The term itself, "analogs", suggests that these products are "not the real thing" and so to be accepted they must convey some advantage to the consumer well above the traditional product. Consumers assume, usually correctly, that they will have to forfeit desirable sensory qualities and so the analog must compensate with extraordinary attributes. Non-dairy cream substitutes conveyed positive value on the dimensions of health (low cholesterol), convenience (storability), and place in the diet (something to go with coffee). Meat analogs have had significant positive worth to some vegetarians, but not to a wider range of consumers.

Cereal blends are directed toward consumers at risk of malnutrition and they have usually been government subsidized. They are not generally appealing to the middle-income families who might afford the commercial product, but they are accepted by needy families when the price is low enough. Taste barriers and nutritional beliefs seem to have been less important than cost in affecting acceptance.

Interpretation

Overall, the two most frequent causes of *rejection* of foods with vegetable protein ingredients have been sensory qualities and cost. We suggest that sensory rejection results from unfulfilled expectations; that is, the image of the traditional food as identified by the consumer was not matched by sensory characteristics. Specifically, consumers expected to taste a familiar meat product and found the taste too discrepant. Inherent unpalatibility seems an unlikely explanation.

Cost has become a cause of rejection when producers have discovered (or anticipated) that consumers perceived no enhanced value in the product containing vegetable proteins that would warrant the higher price. Thus, the intended nutritional appeal of Golden Elbow macaroni or of Incaparina did not persuade perspective consumers to pay more than they would for foods already present in their diets. Underlying this failure lies the fact that the intended consumers did not normally seek nutritional proteins in the pastas or beverages of their usual diets and therefore nutrition was not a perceived inducement.

For those foods that can be considered successes, the causes of consumer *acceptance* are primarily three: price, health, and special functions. Lower price has attracted consumers to those foods in which vegetable protein ingredients have replaced more expensive ingredients -- most have been meat products. The price advantage has worked successfully, however, only in conjunction with the maintenance of sensory qualities and nutritional content of the traditional food. Consumers were unwilling to trade-off a lower price for unfamiliar taste in ground beef in the early 1970's. Maintenance of nutrient content, specifically protein, is also important. The U.S. Army conducted extensive nutrition studies before adopting soy-extended hamburger, and the Japanese might pump hams with starch except for the need to maintain protein content.

The health and nutrition dimension can itself be a selling point for some foods. The notion that proteins are especially "potent", more so than other nutrients, has enhanced the image of several foods, though the idea pertains to the realm of folk belief. More scientifically, nutrition research in recent decades has indicated that consumption of animal fat in Western diets is frequently excessive and so vegetable proteins may sometimes constitute a healthier alternative. Other contemporary, popular health beliefs view grain products and vegetable proteins as generically "healthy". In general, then, foods containing vegetable protein ingredients have benefitted from positive valuation in several health belief systems, some of them more scientifically grounded than others.

Finally, other cases of successful new foods were evaluated positively by consumers because of specialized attributes. The functional properties of soy proteins facilitated reformulation of meat analogs and enhanced the convenience and handling of several foods. These qualities are fairly precise and the foods in which they appear satisfy somewhat narrow markets. For example, the choice of non-dairy cream substitutes and of Incaparina occurs for well-defined reasons. Also, it will be noted, these foods tend not to be reformulations of traditional foods but rather identifiably "new" foods.

All of the new foods reviewed here were untried at the time of their introduction in that they were unlike the foods already known to consumers or they were reformulations of foods known to consumers. There is an important (albeit imprecise) distinction to be made between new foods which consumers perceived to be novel in terms of their prior experience (i.e. "new) and those which were apparently familiar (i.e. reformulations of traditional foods). The distinction corresponds to different marketing approaches and also to different paths of consumer evaluation. For example, the Latin American luncheon meats were nearly a perfect example of a perceived traditional food because no distinguishing information was provided to the consumer; conversely, Puma was clearly perceived as a novel food since it had little resemblance to familiar soft drinks in Guyana. Analogs, by these criteria, are new because consumers have tended to view them as novel. Of the 9 reformulated traditional foods, 6 were considered successes; while only 2 of the 5 new foods seem to have been clear successes.

Foods that are perceived as novel must find some competitive advantage in terms of existing cultural criteria. Conceptual evaluation, as depicted in our model, is predominant in these cases. Call has hypothesized that, "In general, the narrower and more specialized the final market, the greater are the chances of success for a synthetic food or food substitute" (1969). In accordance with his assertion, the few examples of successful new foods among those with vegetable protein ingredients, for example non-dairy creamers, have achieved that position by satisfying particular, and recent, values in the society at large. But cultural values are not easily manipulated and it seems unlikely that meat analogs, for example, will find wide acceptance. Even though attitudes of U.S. consumers may be shifting toward familiarity and acceptance of soy protein, the evidence also indicates that, at present, soy products are not sought as preferred food ingredients.

Instead of attempting to alter consumers' food values, the more recent vegetable protein products have sought to bypass or minimize consumers' conceptual scrutiny by imitating familiar products. Appearing as new ingredients in reformulations of foods that are already accepted has become a promising strategy to get around the conceptual phase of diet selection. The risk, however, is that the imitation must be imperceptible or at least insignificant to the consumer; otherwise the reformulated food is likely to be rejected because it diverges from expectations. This, we believe, is precisely what occurred with hamburger extenders in the early 1970's. It was not that the taste was "bad", but rather that consumers expected it to taste like hamburger and what they ate was too discrepant. The lesson from this experience is that it is better to be distinctively new than to diverge by a little bit.

In general terms, then, two strategies are available to producers. One is to offer distinctively new foods, in which case consumers may be open-minded in their expectations about taste and other sensory attributes. The difficulty will be to make a new food conceptually attractive so that consumers will seek it out. That is, consumers should feel that on at least one value dimension it adds worth to the diet. The success rate for that strategy was low among the foods reviewed here. The second strategy is to replicate familiar foods. In that case, acceptance at the conceptual level is facilitated because it is recognized as a traditional food (and thus already a part of the diet), but the reformulated food must maintain the expected quality of the traditional product or else risk probable rejection at the sensory level.

Conclusions

Food selection in general

1. Human food preferences are influenced by multiple factors operating at the physiological, psychological, and sociocultural levels. Food choices are not directly determined by nutritional need among humans and sociocultural factors predominate over, but do not exclude, the physiological and psychological.

2. Given the opportunity to try new foods, humans both seek novelty and are resistant to change. These opposing inclinations, which may be adaptationally functional for omnivores, lead to moderated trial of new foods. Thus conservatism is only a partial characteristic of human food selection behavior.

3. Humans ascribe values to foods along several dimensions of which nutrition is only one. Therefore perceived nutritional worth may have little influence on the acceptance or rejection of a particular food.

4. In accordance with rational choice, people tend to accept new foods into their diet when they perceive them to enhance the total value of their diet on one or more dimensions. Food values derive from personal experience, but even more from cultural beliefs, attitudes, and symbols.

5. The interplay of general food knowledge and specific sensory qualities is critical to the individual's selection of foods. The conceptual identification of new foods calls forth sensory expectations and, when not met, the probability of rejection is high. Inherent taste preferences are less influential than taste expectations -- discrepant tastes are generally rejected.

6. When presented with a new food, the consumer's perception of whether it is "new" (previously unexperienced) or "traditional" (familiar) influences the degree to which he evaluates it conceptually or sensorially.

313

7. "New" foods receive greater conceptual scrutiny, but sensory evaluation may be less stringent because expectations are less rigidly defined.

8. Foods perceived as traditional can bypass many of the conceptual barriers because they are considered equivalent to foods that are already accepted. However, if they are judged to diverge from expectations at the sensory level they may be quickly reevaluated or rejected.

Selection of foods with vegetable protein ingredients

1. Foods containing vegetable protein ingredients, especially soy products, have been increasingly introduced in the past two decades. Some introductions have been successes and others failures when evaluated in terms of amounts and rates of consumption.

2. Instances of success and failure are distributed across all of the food categories reviewed -- meat products, baked goods, drinks, special markets, analogs, and blends.

3. Of the foods reviewed, a lower proportion of those perceived as "new" were clearly successful. Those "new" foods that were accepted tend to have been tailored to special markets so that they had exceptional appeal to certain consumers.

4. Reformulated foods with vegetable protein ingredients that have maintained traditional quality by matching sensory expectations tend to have been successful. Their acceptance into the diet seems to have been facilitated by consumer familiarity with the traditional product.

5. Reformulated foods that have not met consumers' sensory expectations tend to have been rejected. Taste expectations, and not inherent taste, seem to have been the criterion for rejection.

Appendix A

Physiological And Psychological Theories Of Food Acceptance

Food choices might be influenced biologically in several ways. Preferences are popularly ascribed to "taste" and, it is sometimes suggested, taste and hunger may guide humans and other animals toward those foods that satisfy their nutritional requirements (Jacobs, 1973). The evidence indicates, however, that direct biological influences on human food choice are weak, though not unimportant.

Innate

In the simplest instance, food selection might be governed by taste. The Innate model (Figure 1a) depicts Young's "theory of food acceptance based on the assumption that contact between head receptors and a food object produces an immediate effective arousal" (Young, 1949:103). In that theory, drawing mostly on observations of laboratory rat behavior, foods were selected or rejected on the basis of inherent taste preferences irrespective of nutrient content in the food and nutritional deficits in the animal. Preferences for water and sweet taste were some of the evidence that led Young to conclude that "in general, rats develop drives to run to foods which they *like* (find enjoyable) rather than foods which they *need* (require nutritionally)" (Young, 1949:119). What the Innate model does not explain is how animals alter their diet to meet internal nutritional needs. Preferences are viewed as if they were pre-programmed.

Behavioral Regulation

The Behavioral Regulation model (Figure 1b) is also physiological, but incorporates an influence of nutritional deficiency on food choice. The concept of "specific hungers" was used by Richter to explain results of his studies in the 1930's and 40's on food selection by laboratory rats (1939, 1942). In Richter's experiments the laboratory rats could apparently detect physiological deficiencies and as a result would preferentially choose foods rich in specific nutrients to correct such deficiencies. "Cafeteria" was the term used to describe the experimental apparatus because the animal was allowed to choose freely from a variety of presented foods. On cafeteria dietary regimes, rats were found to be capable of maintaining adequate nutrient intake for an impressive variety of nutrients, including amino acids. Richter therefore proposed a theory of "specific hungers," according to which rats made deficient in some nutrient would develop a specific hunger for foods rich in that nutrient. The precise mechanism by which a rat might detect nutrient rich foods was unknown, but it was assumed that rats must be genetically endowed with specific nutrient receptors capable of identifying nutrients or energy present in potential foods.

315

Models of Individual Food Selection

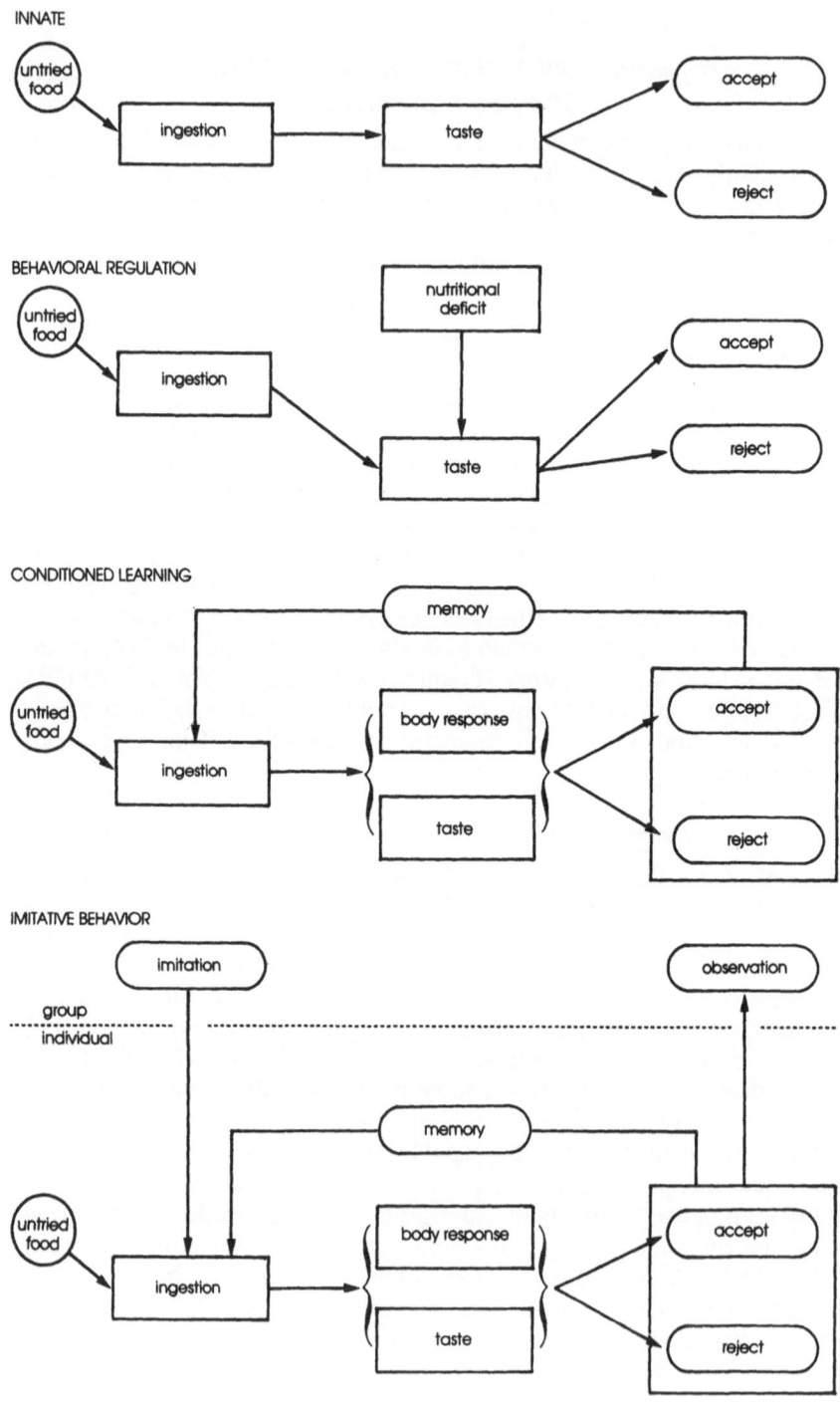

Of the taste receptors, some may be linked to specific hungers at the physiological level. A sensitivity to salt, for example, appears in part to reflect internal sodium balances in rats (Richter 1956; Kaunitz, 1956). Water is another bodily requirement for which there appears to be a direct link between internal supplies and external receptors resulting in the specific drive which we refer to as "thirst." Other taste receptors—bitter, sweet, and sour -- do not appear to service specific nutrient or energy needs. The initial response at least, as suggested by Young, may well be innate. For example, while there is evidence for a distinct preference for "sweet" foods in humans and many other animals (Rozin, 1976), this is a property of a wide variety of foods and cannot be considered a "caloric sensor." Bitter sensitivity, meanwhile, appears to be an important mechanism for detection and rejection of a number of naturally occurring toxins (Garcia and Hawkins, 1975). Most animals respond adversely to bitter taste.

For many mammals, and especially humans, the association between taste perception and nutrient intake seems to be more learned than instinctive. Testing this link experimentally among humans is problematic and, while Clara Davis' self-selection studies of the 1920's and 1930's indicated that infants would freely choose foods to maintain a fairly balanced diet, the possibility of nutritional deficiencies was not built into the experimental design as Richter did later with rats. Even in the absence of experimental results it seems apparent that rats and humans, like other animals, choose among foods by using visual cues and other signals associated with prior experience so that regulated taste must operate only occasionally.

Learning Theories

Dissatisfaction with the limitations of physiological theories led Harris (1933) and Hull (1943) to develop conditioned learning theories of diet choice (Figure 1c). More recently Rozin has proposed a more refined "theory of learned aversions" which is essentially psychological (Rozin, 1976). The emphasis on *individual* learning and choice distinguishes this perspective from the sociocultural mechanisms discussed below. The learning process and the underlying mechanism on which it is based are most clearly evidenced in the rat's well-known ability to learn about poisons and subsequently avoid them. Rats, like humans, are omnivorous and combine certain avoidance behavior, especially of new tastes, "neophobia", with a distinct exploratory tendency, "neophilia". Rat neophobia is potentially so strong that it can be reinforced to the point that experimental animals will starve rather than eat undesirable foods.

Behavioral experiments with rats have shown that food avoidance can be conditioned using a number of stimuli -- bright lights, a buzzer, and smell or flavor -- and a number of induced conditions -- electric shock, nausea, poisoning. Rats learn to associate the stimuli as cues to the undesirable results and hence avoid those foods associated with the stimuli. Even more importantly, taste cues are learned more quickly and have a more enduring

317

effect on behavior than do external cues when associated with foods. Smell and taste seem more closely linked with internal metabolism in the rats' learning process then are visual and auditory cues. Even when nausea is induced 8 to 12 hours after food intake, rats learned to avoid the foods (Garcia, *et.al.*, 1966).

These lines of research help explain the rat's impressive ability to detect and avoid poisons in its environment. Rats, like humans, are essentially conservative in their dietary preferences and look with suspicion on new foods. Moreover, when they do sample new items presented to them, these tend to be consumed initially in small quantities and in discrete "meals" allowing the interval to clearly associate internal consequences with specific stimuli received upon ingestion. If new foods contain poisons, the rat will have learned to associate negative internal consequences with the food and since it consumes only a very small quantity, the rat is unlikely to suffer permanent debiliting consequences. Once a toxic food has been identified, the rat will avoid it in future encounters.

The learned aversions model seems to account well for several aspects of food selection behavior in rats and there are striking similarities with learned avoidance in humans. People's admissions that they no longer eat a certain food because it recalls memories of once having eaten it in excess (or, another type of cue, it recalls an unhappy incident), and the parents urging their children to "just try a little bite" seem to fit well with the theory. The results of a retrospective study of food aversions among 696 subjects, ranging in age from early childhood to old age, suggested learned aversion. Of the sample, 38% had experienced at least one food aversion and, "to an overwhelming degree [87%], taste was the conditioned stimulus and gastrointestinal upset the unconditioned stimulus." (Garb and Stunkard, 1974:1205).

However, the theory of learned food aversion accounts less well for active food seeking and preference. Rozin suggests that in rats, at least, positive food selection occurs through a process of elimination (1976). Given a certain nutritional deficiency, together with the natural neophilic inclination to try new foods, rats may learn to associate improved physiological states with specific foods. Moreover, it seems, exploratory behavior is accentuated by nutritional deficiencies so that the rat begins to turn away from a nutrient-deficient diet and to search for preferable alternatives. That explanation seems to account for Richter's findings of rats deficient in thiamine, when given a choice between foods deficient in thiamine and foods rich in thiamine, shifted their diets to correct internal thiamine imbalances.

There is some evidence that greater preference for certain foods may result from improved physiological states following ingestion. However, rats clearly do not learn as quickly what is good for them as they learn what is bad. While poison food aversions can often be induced after one or two trials, it usually takes much longer for the rat to find nutrient enriched foods under conditions of specific deficiency. For example, when vitamin B deficient rats are offered a choice among new foods, only one of which contains the vitamin, they do not immediately show a selective preference for the enriched food source. Rather, it may be a period of days before a rat samples the enriched source, though only a few eatings of it may suffice to establish a preference (Harris, 1933; Rozin, 1976).

Among humans, the learning of preferences through metabolic response seems even more attenuated. The weight of the evidence indicates that adult humans are, at least, only poorly able to detect and correct nutritional deficiencies by means of these simple learning mechanisms. Indeed, human populations have persistently consumed diets deficient in specific nutrients, as was the case of scurvey among Europeans from the 15th to the 17th centuries, even though nutrient-rich foods were available in their environment. Moreover, clinical cases of nutritional deficiency occur today despite food availability. If human food selection mechanisms were as simple as those described for the rat, such deficiencies might be expected to be corrected quickly.

Imitative Behavior

Recent studies of naturally occurring animal behavior indicate that some feeding behavior may be influenced socially, even among nonhuman primates. A band of Japanese macaques observed over several years adopted an efficient technique to separate wheat kernels from sand after observing the innovation of a young female (Itani, 1973). After trying to pick the wheat out of the sand, she threw a handful into a nearby pond. Apparently the act was fortuitous. The sand sank to the bottom and the kernels floated, enabling her to retrieve the wheat. Through brief trial and error she learned the technique and it was then imitated by others in the band, first by females of the same age, then by younger macaques, and finally by the older males. The pattern of adoption suggested that proximity of status, too, influences imitative learning.

In a separate study, Phillips found that Ethiopian baboons in one region of the Awash River valley tended to consume more leaves and berries of a particular species of tree, *Balanites aegyptiaca*, than other baboons in the vicinity. She inferred, based on several pieces of suggestive evidence, that the plant functioned medicinally to prevent and treat schistosomiasis infection. The incidence of schistosome cercarie and their host snail was shown to be more frequent in the immediate area, infection in the baboon troop and humans was high, and the saponin contained in the *Balanites* fruit and leaves is toxic to the snails and schistosomes. "The co-occurrence of this selective feeding behavior with the incidence of schistosomiasis along the particular stretch of the Awash River, together with the known molluscicidal properties of the *Balanites* berry, supports the hypothesis that baboons are selecting this medicinal plant in those areas where the disease is a threat" (Phillips, 1983: 6). Hypothesizing further, the models of food selection we have reviewed would suggest that one baboon (or a few) might have originally discovered and learned the beneficial qualities of *Balanites* eating in the fashion of conditional learning (Figure 1d). After that initial step, other members of the troop might have learned the practice through observation and imitation.

Examples of the influences of social imitation on food selection are infrequent, yet the importance of imitation as a dominant learning mechanism in human infants would suggest that early food preferences might be conditioned in this fashion. In adults, the acquisition of food preferences is obviously more complex than simple mimicry, however social imitation may still be influential. It seems to be operative when army recruits prefer unfamiliar foods promoted by their sergeant instead of those proferred by a nutritionist in scientist's garb (described below).

Summary

Elements of the food selection models reviewed so far, account for some aspects of human food selection. A tendency to be attracted to sweet and salty flavors and to reject sour and bitter may be due in part to innate preferences. Even though adults frequently override or reverse those initial inclinations, the reaction at first taste, especially among infants, corresponds to primal responses. Evidence for specific hungers is less firm, though it might be exhibited in thirst and certain "cravings" such as geophagia. As a scientific research task, it is difficult to separate learned desires for foods from metabolically induced hungers. Conditioned learning of food preferences and aversions does influence individual diet selection as evidenced above. This mechanism establishes *aversions*, though, more effectively than it promotes.

Appendix B

Explication Of Value Dimensions

The socioeconomic model of food selection suggests that consumers will be more likely to adopt untried foods to which they ascribe higher value on one or more dimensions than foods in their existing diet. Or, in the phrasing of maximization theory, consumers will adopt a food that adds to the perceived value of the diet. We are led, then, to isolate and specify some of the important valuation dimensions in order to test whether they have been responsible for the acceptance or rejection of foods with vegetable protein ingredients. Six dimensions seems especially pertinent: social significance, symbolic and religious meaning, nutrition and health beliefs, place in the diet, functionality and convenience, and price. We examine each of these dimensions below, reviewing studies from several societies to propose how they might be influential in diet change. In the main body of the text we have formulated a brief proposition about the valuation of each value dimension insofar as cross-cultural generalization is possible.

Social Position

The social significance of foods can contribute substantially to adoption or rejection. Certain foods, food styles, or varieties of foods are identified with specific social sectors—an ethnic group, for example, or a socioeconomic stratum, or a regional subpopulation. Foods can also be associated with individual status in society as are baby foods, beer, and tea sandwiches. We have distinguished two aspects of the social dimension in the discussion below—social influences on individual diet choice, and the use of foods as markers of social identity.

Diet is one of the ways by which an individual's status is judged by those around him, and also one of the ways by which individuals seek to manifest their status or aspired status. Much of the research on individual food adoption has been done by psychologists grounded in learning theory. Imitation of peers and peer pressure have been shown experimentally to be effective in molding diet choice (Eppright, 1947). A study of television food advertisements directed at children were shown to be significantly more effective in molding diet when peer-models who were similar to the fourth grade viewers were used to indicate preference (Stoneman and Brady, 1981). A series of studies by the U.S. Army Quartermaster Food and Container Institute on the acceptance of novel foods has been summarized by Peryam (1963). The interpretive stance is clearly that of learning theory:

> Starting at birth, the human infant is exposed to certain practices represented by the kinds of food offered and temporal feeding patterns that are typical of a particular subculture, perhaps as modified by parental idiosyncracies. Until and even beyond adulthood he continues to be exposed to and *indoctrinated with* certain ways of fulfilling his physiological and psychological needs for food instead

of other ways that might be equally rational and feasible. (Peryam, 1963:34., our italics).

Among the findings that supported Peryam's conclusion that social learning was fundamental to food acceptance was that continued exposure or familiarity with several novel experimental foods -- Halvah, chutney, limburger cheese, tripe, and egg foo yung -- did not alter preferences. Moreover, acceptance did vary considerably with the status of the person who introduced the novel food to the army recruits: "the narrator with whom the men should identify more closely (an Army sergeant) was more effective than a food scientist" (Peryam, 1963:37). The use of a role model to influence consumer behavior is well established in commercial advertising and has been used as well in attempts to reverse the trend toward early weaning in developing countries.

Social emulation in diet patterns operates also at the level of social class. Research on food habits in rural, Southern Illinois in the early 1940's conducted by one of us (J.W.B.) concluded that diet patterns distinguished certain classes and could serve as evidence of social mobility: "The general rule seems to be that if one can eat like the group he aspires towards, he has a right to identify with that group" (Bennett, Passin, and Smith, 1943:655). The analysis distinguished three levels of foods -- the core diet, secondary core, and peripheral diet -- on the basis of their frequency of use among several social strata. Diet change seemed to occur as foods were adopted into the peripheral diet and gradually incorporated into the core. Low prestige foods, like river fish, were dropped from the diet as families rose in social status.

Research in the 1940's to 1960's on food habits tended to view diet change as part of the process toward urbanization or assimilation to the mass culture of the larger societies. From that perspective, persistently distinctive ethnic diets appear anomolous. In some instances diet change by ethnic minorities has been reported to follow the general process of assimilation to dominant culture, especially when viewed over generations, but instances of persistence in ethnically distinct diets have also been reported. The research question should be rephrased, perhaps, to be viewed in the context of acculturation more generally and some insights might be gained through closer attention to the differing rates at which components of the diet change. Our review of the evidence below indicates that when social circumstances favor ethnic plurality people may retain some elements of their ethnic diet, even when these elements are not authentically "traditional."

Among the studies that report progressive diet change leading to assimilation with the dominant culture are those on Armenian immigrants in the northeastern U.S. (Nalbanian, Bergan, and Brown, 1981), three generations of Japanese immigrants to Hawaii (Wenknam and Wolff, 1970), and Asian Indian immigrants to the United States (Gupta, 1965). Consistent with acculturation theory the Japanese study concludes that:

From a predominantly high carbohydrate, rice and plant food diet, the pattern of the Japanese immigrant family in Hawaii has changed to a high-protein and fat, "cosmopolitan" diet. Among the socio-cultural factors which brought this change about are: desire for higher social status through adoption of a diet associated with the higher status group at the same time traditional family controls were eroding (Wenknam and Wolff, 1970: 32).

The Indian study notes that, "the food habits of Indians become increasingly Westernized" (Gupta, 1965: 98) and the Armenian study concludes that "the characteristics of an ethnic diet gradually disappear along with other cultural trademarks" (Nalbanian, Bergen, and Brown, 1981: 698). Despite their conclusions, however, some components of the diet had changed more quickly than others in all three cases, and they were still distinct in some ways from the predominant diet. Among all three, breakfast had acculturated most quickly and the main meal least changed. Traditional meals had changed most in terms of ingredients and form, but least in regard to seasoning and method of preparation. The third generation Armenians, for example, continued to make a meal they termed *losh-kebab*, even though such vast changes in the form had occurred that the traditional food was hardly recognizable in the new.

A study in Madras, India on the descendents of immigrants from the northern state of Maharastra emphasized the retention of some dietary habits as markers of social differentiation. Katona-Apte (1975) states that, despite superficial appearances, the Marathi immigrants retained a distinct ethnic identity.

> To the eye of an outsider, the Marathi speakers appear to have assimilated considerably into like in South India. When questioned, they claim to be South Indians. They speak Tamil as any native, dress as other South Indians, and seem to eat South Indian foods. But in the home and among relatives they speak Marathi, they marry other Marathi speakers only from South India, and they celebrate Maharashtrian holidays and festivals. And on closer examination, their food habits do differ from those of other South Indians (1975: 322).

Katona-Apte found that diet assimilation was most advanced among the Marathi speakers in those aspects of the diet which were similar or "suitable" to the dominant culture and in the daily diet. The least-changed aspects were the tradititional habits of food preparation and consumption which accompanied Maharashtrian festival celebrations.

The overriding point to be made was well-stated by McKenzie in a similar study on minority diets in Great Britain, "food habits set the group apart from the community as a whole" (1967: 201). Under prejudicial conditions a minority group may prefer not to be "set apart" and so diet patterns like other ethnic markers, may be abandoned fairly quickly. On the other hand, where social conditions tolerate or even favor ethnic pluralism, then some food customs may be retained, even though the "tradition" may not, in fact, be authentic. Those diet components that seem to have greatest retention potential are flavoring, preparation style, and meals with special social significance. The "main meal" and festive meals have particular longevity probably because they are more central to social events in the activities of a family or community. The U.S., it has been suggested (Glazer and Moynihan, 1963) has progressed "beyond the melting pot" and ethnic pluralism is fostered in many settings so that dietary assimilation can no longer be viewed as a one-way street.

Price

Ideally, in a market economy, price should be the common denominator of all value in goods. It is not surprising, therefore, that price is a frequent determinant of initial attraction for many introduced foods. Most simply stated, consumers will prefer a less expensive good, all other things being equal, to one that is more expensive. Notorious exceptions, like a premium price paid for "the label" of a wine, for example, do not disprove this principle, but rather confirm that some of the values ascribed to goods are intangible.

Two aspects of the pricing dimension are particularly relevant to foods with vegetable protein ingredients. They are (1) the disproportionate frequency with which price appears as a determining factor in the acceptance of meat products (see Table 2) and (2) the provisioning of foods through non-price, non-commercial channels.

meat products: As a protein source, soy products generally have a considerable cost advantage relative to meat and this has been a prime motive for attempting to introduce them into the place of meat in the usual diet. In general, consumer demand for food rises more slowly than the demand for other goods as income rises. Ernst Engel established this economic principle with his study of Belgian family diets in the mid-19th century. The income elasticity of some foods, though, is much higher than others. Diet staples, or "common fare", which in many societies are cereal based, tend to vary little with income. Meat consumption, on the other hand, is closely correlated with income both within societies and across nations (Pyke, 1978).

Consumption of meat, especially beef, has gradually assumed prominence in U.S. diets and from this trendsetting base it now influences diet patterns in the rest of the world. Societies which have been traditionally meat scarce now experience increasing demand from those who can afford it. Pyke writes (1978), "The persistent popularity of beef among communities which, one after the other, have embraced the technology of the West, is remarkable in view of the extravagance represented by the inefficiency with which a steer converts the protein it eats into meat protein for human consumption."

The esteemed place of meat in the Western diet goes back at least as far as the 11th and 14th century when European meals comprising several meat plates --both wild game and domesticated animals -- were usual even among the common classes (Braudel,1973: 130). In the 15th to 17th centuries meat was

324

less available but its desirability was, if anything, enhanced. Meat consumption, especially beef in England, was a sign of prestige. The European meal structure, with meat as a centerpiece, was transported to North America where it has taken on a trajectory of its own. The grazing of cattle on the midwestern plains and the establishment of refrigerated railway shipping of beef to Eastern cities in th 1860's by men like Philip Armor and Gustavus Swift expanded the market for beef (Harris and Ross, 1978). In the 1950's beef surpassed pork as the major meat source in U.S. diets and by the mid-1970's over 40% of beef was consumed as ground beef.

The preference for meat and its relatively high price make it sensitive to price shifts and susceptible to less expensive alternatives. In recent years, U.S. consumers have shifted between beef, pork, and poultry as prices have changed. Nutritional considerations have also influenced these trends. Beef consumption fell from a peak of 91.8 pounds *per capita* in 1977 to 77.1 pounds *per capita* in 1981 *(Wall Street Journal, 4/1/82)*. Another significant trend has been toward less expensive cuts of beef, much of it from cows instead of grain fed beef (New York Times, 1/7/81). It seems likely that the increased world demand for meat coinciding with the reduced demand for prime quality beef will create favorable conditions for extended meat products. The price differential between meat protein and vegetable protein seems to lead inevitably toward that projection although, it should be noted, similar predictions made in the mid-1970's (Minor and Gallimore, 1977) have been proven to have been overestimates.

non-commercial channels: In many parts of the world, cash purchase has only recently become an important means for obtaining food. In tribal and peasant economies, food provisioning is traditionally embedded in longstanding social structural arrangements. Trading partners may reciprocally exchange food and other goods over years, or food surplusses may be accumulated and redistributed through the community in some ritual fashion. Anthropologists have viewed these as alternative, non-market economic systems for provisioning societies above the household level. These arrangements, it has been argued, enable the society to support individuals or households at times of occasional need. The incursions of a cash economy can, it is argued (Ortiz, 1973), disrupt functioning exchange patterns and lead to impoverishment.

Field research in several societies has documented the shift from subsistence based agrarian economies to cash income and food purchasing. In nutritional terms, the shift has sometimes had negative consequences as, for example, among Brazilian sisal laborers studied by Gross and Underwood (1971). But positive nutritional effects have also been reported in Papua, New Guinea (Harvey and Heywood, 1983) and Costa Rica (Sellers, 1984). Sellers showed that, under prevailing economic conditions in Costa Rica, the optimal strategy for family farmers was to raise crops for market sale while purchasing their foodstuffs. Subsistence farming was generally uneconomical.

The threat to social welfare, as sectors of societies move from local food provisioning to commercial exchange, is that individuals become more reliant or dependent on the system for their well-being. They become highly susceptible to increases or drops in food cost or unemployment. For this reason, nutritional problems are often more severe in the marginal urban neighborhoods of cities in developing countries than in the rural sector (Den Hartog, 1981). As the market system has permeated the economies of developing societies, malnutrition has become increasingly a consequence of poverty; that is, individuals are unable to *purchase* a minimally adequate diet.

One solution used to address malnutrition under these circumstances is to bypass the market system by providing food relief either directly to needy consumers or through food substitutes.

Nutrition Beliefs

From a purely functional viewpoint, nutritional value might be considered the quintessential criterion for food selection. The finding from behavioral research, though, is that nutritional content is only one of several value dimensions on which foods are judged. It is not insignificant, however, and must be considered one of the influential dimensions. The major point to be made in the following discussion is that perceived nutritional value of any food is derived from a food (or health) belief system. Whether people's nutritional beliefs are scientifically valid or not has little consequence for their food choices (although, of course, it may have important consequences for their well-being). In the U.S., popular nutrition beliefs have gradually approximated scientific understanding in recent decades through the efforts of educators, scientists, popular new media, and government institutions, among others. Still, the correspondence is not perfect and so an understanding of nutritional beliefs apart from the scientific understanding of nutrition is essential.

The following two examples of diet health belief systems from Mexico and Australia, show how foods are chosen for reasons that follow the logic of the belief systems. They also exemplify another principle of behavioral analysis -- that the connection between symbol and meaning is frequently arbitrary. It will be seen that both belief systems base themselves on a hot-cold dichotomy but they attribute opposite meaning to concepts.

The traditional subsistence system among Australian aborigines is hunting and gathering. Kangaroos are the major source of meat; other marsupials and anteaters are also hunted. Foods that are gathered include fruit, seeds, tubers, and leaves. An informant who was asked to describe "how the body works", said to the ethnographer:

> We Aborigines say *mulgurba*, that is, hungry; a fellow becomes hungry and hot, he becomes hungry and hot without meat, he becomes overheated. Being hungry and without meat he would go out quickly for bush-tucker. He goes away overheated and hungry. He eats and becomes full only by picking up and eating a great amount of

bush tucker. He eats and becomes full and this makes his heart cold, this makes his head cold, this makes his blood cold . . . Meat makes the head full and the spirit full, meat gives blood right up to the head. Blood goes to the eyes, the two ears, and both thighs. I would die if I had no meat or water. Meat keeps up my blood, bush-tucker also gives me blood. (Winninydji and A.R. Peile, 1978: 497-523)

According to Aboriginal beliefs hot and dry are unhealthy conditions, while cold and moist are healthy. "Health is derived from blood, for blood is life and the source of this life is primarily meat" (p. 504). These beliefs have ramifications in food exchange, ritual, and healing practices so that a shortage of meat threatens many aspects of Aboriginal well-being. Newly introduced foods or medicines would be expected to gain wide acceptance if they connoted the positive qualities of cool and moist.

It seems unlikely that there was any historical link between Australian beliefs and the hot-cold system of Latin America. This second system has its origin in Greek humoral pathology which, elaborated by the Roman Galen and Arab physicians, reached Spain by way of Moslem occupation during the 14th and 15th centuries. "According to this system each of the four humors had its 'complexion', consisting of pairs of qualities. Blood was hot and moist; phlegm, cold and moist; black bile, cold and dry; and yellow bile, hot and dry" (Foster, 1967: 185). What was then accepted medical opinion has filtered down to the common culture and constitutes the basis of popular beliefs about food and nutrition throughout Latin America. Some foods are popularly categorized as hot (regardless of serving temperature or cooking preparation) while others are cold. Specific assignment of foods to categories varies regionally and even locally (another instance of arbitrary symbol-meaning associations); in Michoacan, Mexico hot items included beef, pork, ice (!), wheat, beans, coffee, and tobacco while some cold items were chicken, beer, rice, maize, potatoes, and oranges (Foster, 1967: 187).

The hot-cold categories are used in Latin America to prescribe or proscribe certain foods in the diet contingent on other body circumstances. In the normal, healthy state the body is thought to be at an equilibrium between hot and cold. Certain activities, like emotional excitement or exposure to drafts, are believed to heat or cool the body (correspondingly for these two examples) and those conditions should not be exacerbated by eating hot or cold foods. The underlying objective is to avoid extremes and to maintain a "normal" equilibrium. It is not uncommon to find that in actual practice people disagree about whether a particular food will aggravate or restore an imbalance and the belief system is most successfully employed *ex post facto*2 to rationalize a discomfort or illness (*i.e.,* "It must have been something I ate.").

What these two hot-cold belief systems exemplify is the function of folk theories of diet and health. Like cultural belief systems in general, they make peoples' activities intelligible to themselves. These two particular systems build on the hot-cold dichotomy which, like other naturally occurring dichotomies (left-right, dark-light, male-female) provides a convenient binomial base for defining correct versus incorrect behavior. The fact that they assign different values and meanings to the hot-cold distinction is further illustration of the frequent arbitrary assignment of function to form in cultural belief systems.

In the U.S., the scientific understanding of nutrition is significantly different from these folk dietaries in that it seeks to be grounded on objective, empirical evidence and hypothesis testing. This should not lead us to overlook, however, the substantial body of popular nutritional beliefs that are also part of our culture and often at odds with scientific knowledge. Americans tend to eat what they think is good for them (though not even this, always), but information is drawn from a variety of sources.

Commenting on U.S. food habits one perplexed researcher lamented in 1947 that "The attitude of people toward health and food is difficult to understand. The public is health minded. It is apparently convinced that certain constituents of foods, the vitamins in particular, promote health...Nevertheless, urging greater use of certain foods because of their health-giving properties seems not to make a popular appeal." (Eppright, 1947: 89584). Vitamin consumption and the general appeal of "health foods" are even greater now in the U.S. than in 1947. The imaginative titles of more recent works by respected nutritionists "American's Love Hogwash" and Deutsch's "Nuts among the Berries" are testimony to scientists' frustration with the public's gullibility for health food fads.

Several explanations of American attitudes toward health foods have been proposed. One is that it reflects a social psychological obsession: "We would strongly suggest that health food peddlers capitalize on one of the fetishes which is very marketable in our American society today: beauty and youth" (New and Priest, 1967). An extension of this explanation is it is the symbolic value of the health foods, how the consumer uses them to lend meaning to a desired lifestyle, that is most significant. The symbolic potency of certain foods, the ritual avoidance of "unhealthy" foods, and the special preparation often required suggest that something more than just eating is involved. Another explanation stresses consumer ignorance and gullibility: "Most of these faddists should be considered as psychologically exploited rather than suffering from some neuroses" (Todhunter, 1973: 306). Examples of persuasive quackery seem to substantiate this interpretation, but they do not account for much of the apparently voluntary selection of health foods. We suggest that there is some truth in both explanations but that they might be complemented by a third: they are fringe offshoots of an honorable tradition in American culture -- a concern for nutritional health and a suspicion of "artificial" foods. That tradition has sponsored inventions like corn flakes, Graham cracker, and peanut butter and it lies behind public skepticism of chemical food additives. Balanced skepticism and "health mindedness" are probably advantageous to society and individuals; health food fads may be an extreme offshoot of that tradition.

Another manifestation of this attenuated nutritional view in the U.S. is the acute suspicion of adulterated foods. In popular magazine articles, soy analogs have been referred to as "unfoods" (Sanford, 1968: 13) and "fake" foods (Kent, 1977: 20). One reported alarmingly that "there has been no resistance at the supermarket, even to some of the older synthetics -- such as *Postum* and oleomargarine -- since the turn of the century" (Kent, 1977: 20). These views are evidently more sensationalistic than the cultural norm, but they do draw on a common American sentiment that processed food is "unnatural" and unhealthy.

Pyke makes the point that, compared to most societies, U.S. diet selection corresponds fairly closely to scientific understanding: "Americans have attached more importance than Europeans to the contribution their diet makes to health and nutrition and have been quickest to change their diet in relation to what they have thought to be the dictates of nutritional and medical science" (1975: 90). Several trends in the changing U.S. diet over the past three decades are revealed in statistical overviews and can be attributed to the impact of nutrition research on public attitudes (McCann and Trulson, 1957; Page and Friend, 1978; Winikoff, 1978; Schwerin, et al, 1981).

An analysis in 1957 of diet change in the U.S. over the first half of the century concluded that "The major changes in the American diet appear to be a decrease in the consumption of total calories, a marked decrease in intake of carbohydrates, particularly from grain and cereal sources, with a concurrent increase in consumption of fat and an increase in the use of hydrogenated fat" (McCann and Trulson, 1957: 363). At the time of that writing, research had just begun to publicize the likely effects of saturated fats on atherosclerosis and heart disease. Reviews 20 years later revealed the same longterm trends but the latest data indicated some shifts away from animal fats and sugar toward mixed protein sources and high-fiber grain products. Research on diet, coronary heart disease, blood pressure and intestinal cancers probably influenced these trends. One review tentatively concluded that "the trend is moderately encouraging. At the same time that they have been consuming a little less food by weight, people show some signs of moving at best toward better combinations of the foods they are eating" (Schwerin, *et al.*, 1981: 56).

Foods for malnutrition assistance, like the cereal blends are usually government supported and cannot be judged by strict market criteria. Indeed, they are out of balance with demand-supply determined prices as their erstwhile commercial competitors have unhappily discovered. Nutritional impact is a primary consideration for such products and so it becomes necessary to evaluate improvement or prevention of malnutrition in populations at risk. For example, the decision not to implement amino acid fortification of grains was made by the U.S. assistance agencies after field studies with fortified wheat, rice, and corn failed to demonstrate significant impact (Austin and Zeitlin, 1981). Cereal blends, like corn soy milk, have had demonstrable effects on malnutrition, but having passed that first criterion they must then be considered relative to alternative nutrition interventions in terms of cost-effectiveness, availability of resources, duration of impact, implementation feasibility, etc.

In general it can be said that protein-energy malnutrition in the world today is best addressed with inexpensive cereal blends. Recent research indicates that protein is not so critically lacking as was once thought and a more comprehensive, energy replete diet, is optimal. The less expensive grades of soy will probably continue to be prominent ingredients in such blends because of soya's balanced amino acid content, but cost relative to other protein source will also determine its use. Another consideration for many governments in developing nations is the cost of importing vegetable protein ingredients. Balance of trade deficits have led to the discontinuation of imported soy ingredients in several countries and the solution could be to turn to locally produced vegetable protein sources. The lack of commercial viability for fortified blends and the infrequency of local government subsidies make this improbable, however.

More attention might be given to cooperative arrangements between government and industry in addressing nutritional deficiencies. Because public health budgets are limited, especially in developing countries, food subsidies are insufficient to alleviate all nutritional problems. On the other hand, when subsidies are applied they tend to displace commercial alternatives. If ingredients of acceptable or appealing diet supplements were identified for different socioeconomic strata, then governments might facilitate commercial entry into those markets without incurring major budgetary outlays. At the lowest poverty levels, heavy subsidization may be necessary, but at slightly higher levels commercial markets for inexpensive, nutritious foods can be identified. Vegetable protein ingredients might easily find a place in those markets.

Symbolic and Religious Value

Foods are so frequently ascribed symbolic value that it is impossible to review the topic comprehensively. What we do here is illustrate some outstanding principles of food symbolism which are relevant to the acceptance of new foods.

The symbolic value of some foods supercedes any other valuation. Communion bread in the Catholic mass or pork to a Muslim are evaluated in terms of religious precepts to the exclusion of everything else. Other foods may be symbolically "charged" to a lesser degree, either negatively proscribed ("taboo") or positively prescribed. Since our focus is on food acceptance, we have treated the symbolic dimension as it were a "charge" or attribute associated potentially with any food. However, symbolic food values generally derive from integrated belief systems and so interpretation must look beyond individual foods to the larger system in which they are embedded.

330

Cannibalism, the Jewish Kosher dietary, and the Hindu prohibition on beef are among the many food habits that are enigmatic at a first nutritional glance. Nutritionists are also perplexed by apparently dysfunctinal food taboos. A researcher in Nigeria, examining food beliefs and nutrition, found that meat, eggs, and oil, among other things, were tabooed for young children and might have alleviated some malnutrition." "The extent to which these food restrictions affect the nutritional status of a community is often not fully realized. The forbidden foods could play an important part in the diet of an unsophisticated people" (Ogbeide, 1974: 215). The apparent nutritional dysfunction of some food taboos leads us to ask what purpose they might serve.

Several kinds of explanations for food symbolism and taboos have been proposed. We summarize 6 of them here and then propose some general principles of the symbolic dimension. The 6 explanations are: 1) health and sanitation, 2) ecology, 3) ethnic and social identity, 4) cognitive structure, 5) acts of piety or moral worth, 6) communication (see Grivetti and Pangborn, 1974 for a similar list).

1. health and sanitation: The Old Testament prohibitions against eating some animals (camel, pig, crow, some locusts, lizards, etc. Leviticus, xi), have popularly been viewed as codified hygienic measures. Many of the animals, it is argued, might carry infectious disease, spoil easily, or transmit parasites. On closer inspection, though, the argument is generally unconvincing. The Hebrew distinction between clean and unclean does not correspond to real or perceived animal hygiene and, moreover, many of the same animals were consumed by nearby tribes (Douglas, 1966; Simoons, 1978).

2. ecology: Several scientists have argued that food beliefs, no matter how peculiar, must serve some positive adaptive function for the population. Veneration of the pig among New Guineans, for example, has been seen as a means to maintain availability of animal protein in balance with the human population (Rappaport, 1968). The opposite custom, pig taboo, has also been explained ecologically. Since pigs eat many of the same foods that humans do, they may be direct competitors for grain, tubers, nuts, and fruit. As populations increased and forests diminished in the Middle East and Mediterranean, humans excluded pigs from their shared ecological niche by prohibiting them with taboos (Coon, 1951; Harris, 1974). Similarly, Hindu veneration of the cow and the prohibition of beef consumption have been interpreted ecologically. The argument is that live cows supply more (milk, dung for cooking, etc.) to the Indian food system than they would if slaughtered and consumed (Harris, 1977). An extreme application of the ecological approach has been to account for human sacrifice and cannibalism among the Aztec. Briefly stated, the argument is that ritualized cannibalism supplied the Aztec state with scarce meat -- there being no large, domesticated livestock in Middle America. "From the Aztec point of view, the Tlaxcalan state was preserved as a stockyard." (Harner, 1977: 51).

One failing of an ecological explanation for food symbolism is that it seems capable of accounting for almost anything, but always in an *ex post facto* fashion. It imputes a rationale to customs which the practitioners -- the Hindus or Moslems -- do not themselves acknowledge and its working assumption, that all customs *must have* some adaptive survival function, requires empirical foundation beyond blind faith. Why, for example, should the Aztec need for meat protein have led them to cannibalism while meat-scarce India is largely vegetarian?

The ecological perspective is most usefully viewed as one of several possible explanations for food symbolism -- and one which can be convincingly substantiated in some instances.

3. ethnic and social identity: We have pointed out elsewhere that foods can be used to mark social boundaries; that is, in the terminology of the National Research Council, foods "edible by human beings, but not my kind of human beings" (Lowenberg, 1970). Some religious prohibitions may be liturgical means to define and reinforce the distinction of in-group from out-group.

Jewish dietary prohibitions in the Old Testament have been ascribed the function of fomenting social cohesion among the Jews by accentuating the differences in their food habits and those of their neighbors (Korff 1966). Specifically, they may have sought to distinguish themselves as pastoralists from neighboring nomads and hunters: "Cloven hoofed. and chewing ungulates are the model of the proper kind of food for pastoralists. If they must eat wild game, they can eat wild game that shares these distinctive characteristics and is therefore of the same general species" (Douglas 1966: 54). The Jewish prohibition on pork may have intentionally set them apart from the Egyptians and Babylonians who dominated them at different times. The Babylonians not only consumed pork but also used pigs in ceremonial sacrifice. Centuries later, Catholic Spain reversed the tables by consuming pork in contradistinction to Jews and Moslems. "The theme of ethnic identity was carried to extremes during the Spanish Inquisition, with the conspicuous consumption of pork and the suspension of hams from the ceiling over dining tables as visual testament of not being Jewish or Moslem" (Grivetti and Pangborn, 1974: 636).

Social identity will not account for all food symbolism, but does seem prominent in many instances. The observance of religious dietaries by Orthodox Jews, Seventh-Day Adventists, and Black Muslims, among others seems, at least in part, to reinforce social cohesion. The emphasis is on avoidance rather than prescription, and the prohibited foods are prominent in the dominant, surrounding culture thus accentuating "apartness."

4. piety and moral worth: The preceding rubric referred to social function, while the present speaks to individual behavior based on the same principle. Religious proscription for the group implies abstinence for the individual, and abstinence can be considered an act of piety or moral restraint. Orthodox Jews who maintain the kosher dietary sometimes compare it to a Catholic's telling the rosary; that is, small acts serve as reminders of the individual's relationship to a deity. Self-sacrifice evidences and amplifies devotion.

Ceremonial fasting is a more generalized expression of abstinence. For example, members of the Ethiopian Orthodox Church frequently learn fasting as children and, as adults, may fast for 1/3 to 1/2 of the days of the year (Knutsson and Selinus, 1970). The practice is understood as part of a world-view which contrasts mental with physical, mind with stomach, knowledge with food, divine inspiration with human emotions and aggression. "Through fasting man attempts to minimize the evil stemming from human aggres-siveness, power, and pride, and, at the same time, he is bridging the cleavage between two parts of reality, that of human weakness and divine might" (Knuttson and Selinus, 1970: 967).

For the individual, the attitude that food avoidance entails shunning improper or profane foods has transcendent implications for the entire lifestyle. Studies among "new vegetarians" in the U.S. found that 37% "claimed that changes were in outlook and state of mind. They described themselves as 'calmer', 'more peaceful', 'more stable', 'less materialistic', and 'less chaotic and disorganized'. (Dwyer, *et al.*, 1974: 380). None of these new vegetarians belonged to the major vegetarian religious sects, though many viewed their vegetarianism in moral or religious terms.

Formal religious adherence tends to encourage and reinforce acts of piety. As an act of self-denial, dietary proscription makes most sense in the midst of available, even desirable, foods. That is, abstinence has little significance if the foods are not available anyway. For this reason, the nonmeat "bacon", "sausages", and "hamburgers" consumed by Seventh-Day Adventists are understandable—they are obvious symbols of self-denial.

5. cognitive: Beginning with the notion that culture provides the schema by which the world made intelligible to humans, some anthropologists have seen taboos as "cognitive anomalies". The idea is that things do not properly fit in their expected taxa, or fall across conceptual interstices, are improper and, by extension, inedible. Says one advocate of this view, "The general theory is that taboo applies to categories which are anomolous with respect to clear-cut category oppositions" (Leach, 1964: 39).

Douglas has applied this idea imaginatively to some of the Old Testament proscriptions:

> In general the underlying principle of cleanness in animals is that they
> shall conform fully to their class...To grasp this scheme we need to go
> back to Genesis and the creation. Here a three-fold classification
> unfolds, divided between the earth, the waters, and the firmament.
> Leviticus takes up this scheme and allots to each element its proper
> kind of animal life. In the firmament two-legged fowls fly with wings.
> In the water scaly fish swim with fins. On the earth, four-legged
> animals hop, jump or walk. Any class of creation which is not

equipped for the right kind of locomotion in its element is contrary to holiness. (Douglas, 1966: 55)

This principle, Douglas holds, accounts for the taboos on the crocodile, chameleon, mole, and eel, among others.

Leach used a similar scheme, but based on categories of perceived social distance from man, to account for prescriptions on eating dogs, horses, and insects (Leach, 1964) though objections have been raised to the logic of the presentation (Halverson, 1977). Tambiah has elaborated Douglas' principle to a set of hypothesized corollaries, for example "An unaffiliated animal, if it is seen as capable of leaving its location or habitat and invading a location or habitat of primary value to men, will be the focus of strong attitudes expressed in the forms of (1) a food taboo and (2) a bad omen or inauspicious sign" (Tambiah, 1969: 450).

The principle of cognitive anomolies of taxonomic dissonance does not apply to all instance of food symbolism, but does seem to explain some that are otherwise enigmatic.

6. communication: Because food can be endowed with symbolic value, some dietary practices seem to bear most significance as acts of communication; that is, the food conveys special meaning to those "in the know."

The communion elements -- bread and wine -- from the Catholic mass are foods used metaphorically. To the participants their meaning transcends their material substance through consecration, and their ingestion places the communicants in contact with the holy. Layers of meaning are lent to the act of eating.

But symbolic meaning is culturally endowed. "Sinhalese villagers do not like being watched while eating. They will almost always cook indoors. In the villages they will also eat indoors with the shutters or windows firmly closed." (Yalman, 1969: 81). In analyzing their food habits, Yalman decides that eating has special symbolic significance for the Sinhalese. First, it implies intimacy, as sexual intimacy between a husband and wife. Second, because of the rules of caste pollution, foods should not be received from those of inferior caste. And third, food offerings to Buddha and other deities substitute animal sacrifices and suggest equality and communication between man and God. In short, the act of eating among Sinhalese conveys many potential messages and so is not a casual, public act.

It seems likely that Aztec human sacrifice and cannibalism was also largely a metaphorical act. The priests' wearing the victims' skins and the ritualized consumption and sharing of parts of corpse symbolize identification with the victim and the sacrifice. At the same time, it was a consummate message of domination and absorption by the Aztec empire over subjugated tribes.

Summary: No single principle accounts for the symbolic value ascribed to foods. The six described here seem operative in certain instances and they may overlap or work in tandem. Because symbolic content is culturally ascribed, few generalizations are possible; rather specific study is required. Two trends, however, do seem to prevail across cultures: (1) food proscription is more frequent than prescription and (2) taboos usually concern meat and animal products rather than plants. Simoons says, "Flesh foods tend to present far more serious problems in attempts to gain cultural acceptance than do most foods of plant origin" (Simoons, 1978: 178).

Function and Convenience

The most pervasive way in which the U.S. has set the trend for diet changes in the world is in the technological enhancement of functionality and convenience. One European commentator says, with a notable lack of adulation,

> One of the effects American food technology has had on the character of European society has been to accelerate the extinction of the general store on the street corner, of the specialized butcher, baker, greengrocer, and dairy and to substitute the supermarket...As the leader in prepackaging, canning, freezing...the United States could be said to have played some part in the change in the social coherence of the family, for which advances in food technology have in some measure been responsible...Convenience foods have not only abbreviated the preparations needed for a family meal, they have made such a meal a dietetic irrelevance. People can now prepare and consume prepackaged commodities individually. (Pyke, 1975: 90)

The impulses to transform diet technologically have been several. Consumer demand for food variety seems to have played a role in promoting innovations in transportation and storage. In the second half of the 19th century in the U.S., the intercontinental railroad, canning, and freezing, together with mechanization of grain production began to supply a bountiful supply of meats, fruits and vegetables to urban centers throughout the year. "The new middle classes -- possessing no landed estates which could supply them with the material for their *entrees* and *entremets* -- were in the market for as wide a variety of socially acceptable foods as the world could provide." (Tannahill, 1973: 347). By the turn of the century, fruits were being shipped in quantity from the tropics and fresh vegetables reached East Coast cities from California's Imperial Valley.

The second impulse, accelerating in importance since the Industrial Revolution, is the desire to reduce time spent in food procurement, preparation, and consumption. Work at the office or factory reduced family meals, but even more significantly, industrialization gave a new meaning to time. "Time is money," said Benjamin Franklin in a characteristic aphorism. (Franklin, was the quintessential capitalist entrepreneur in the opinion of sociologist Max Weber.) Many diet changes of the past century reflect less time spent on food: the increase in processed fruits (especially orange juice) and vegetables (dehydrated potatoes), the increase in sugar (soft drinks and snacks), the increase in ground beef (hamburgers), and the prominence of "fast food" restaurants. (Page and Friend, 1978; Hollingsworth, 1961). Comparing English diets in the 1960's with those a century earlier, Barker concludes that the changes have been toward greater labor saving and palatability but not higher nutritive content. (Barker, 1978: 175).

The third impulse is just now emerging in the U.S. and is largely unperceived: an intentional devotion of time to selected meals. The faddish interest in "gourmet" cooking or dining is one aspect of the phenomenon, but it should be recognized as functionally coupled with convenience foods. "Schizophrenia characterizes the institution of eating. At one time we will frequent the highly symbolized yet sterile world of fast food, while at another we will leisurely partake of the ritualistic world of gourmet eating and formal dinners." (Curry and Jiobu, 1980: 253). The observation here is accurate, but the characterization is wrong; rather than "schizophrenia" what we are witnessing is an intentional weighing of eating versus other activities. Contemporary technology has brought middle class Americans to the point of meaningful choice among activities: dining as a pleasurable, time-consuming activity versus, for example, a quick meal and a movie.

The functional dimension, then, may have two positive values. On one, features that provide time-saving, convenience, improved storage, etc. will be preferred. On the other, foods that enhance the usefulness of foods as "gourmet" style or for special occasions may be positively valued.

Place in the Diet

Diets are structured culturally in several ways so that if something is "out of place" it may be considered incongruous, inedible, or undesirable even though it might be readily consumed in another context. Thus, spaghetti for breakfast or strawberry jelly to accompany fried shrimp would be unappealing to most Americans. Place in the diet is an important dimension for evaluating a new food. It has been hypothesized that foods which fit a familiar diet niche will have a higher likelihood of acceptance than those which do not fit.

structure of cuisine: A useful way to compare food styles across cultures is by 3 dimensions: the basic ingredients, preparation methods, and flavoring (E. Rozin, 1982). The basic foods are often distinctive and characteristic of a culture's cuisine. Corn and beans in Mexico and Central America are diet staples and typify the diet. Rice is characteristic of all of Asia and there are regional variants: rice, fish, and coconuts in Malaysia (Wilson, 1975), mushrooms, soybeans, and snow peas in China (Williams, 1973), fish and alga in Japan. Around the Mediterranean and in the Middle East, legumes and breads made from several grains typify different regional patterns (Kariel, 1966). In the Greek diet, for example, wheat bread is central to the meal and other foods complement the bread (Williams, 1973: 281).

A second dimension of cuisine, preparation method, can be broken down further into processing to change the form of the ingredients (chopping, slicing, whipping, separation, etc.), processing to cure or manipulate water content (marination, pickling, drying, freezing, etc.), and cooking (baking, boiling, frying, etc.) (E. Rozin, 1982). All cultures seem to have rules for food preparation (i.e., "recipes") and cooking is a cultural universal, even though it is not a biological necessity. In some cuisines, though, the method of preparation seems especially distinctive. Stir-frying is characteristic of Chinese and Japanese cooking. Pit baking with heated stones is associated especially with Oceania.

Flavoring and seasoning is a third culinary dimension and perhaps the most salient in discriminating cultural food styles. Soy sauce is distinctively Oriental, curry mixes are identifiable with India and Southeast Asia, sauces of hot pepper and tomato are Mexican. The flavor may be so definitive that ingredients and mode of preparation become inconsequential. "Cover any food, no matter what, with a sauce made of tomatoes, olive oil, garlic and herbs and we identify it as Italian; what is more, Italians will identify it as Italian. Be it dromedary hump or acorn meal, its culinary identification will ultimately be determined by the way in which it is flavored." (E. Rozin, 1982: 197). Some flavorings are so peculiar that nonmembers of the culture are repulsed: the blend of chocolate and piquante in Mexican *mole*, for example, is unappealing to most Americans.

meal pattern: Another way in which food behavior is patterned is through meals. Most Western diets vary over a daily meal cycle, a weekly cycle, and an annual cycle. Meals occurring in the annual cycle accompany calendrical celebrations, many of them associated with civil holidays and religious festivities. Turkey and cranberry are identified with the Thanksgiving meal in the U.S., hot cross buns with Ash Wednesday. The weekly cycle coincides with the work pattern and some religious observances: Sunday dinner as a special meal, for example. The daily cycle is generally thought of in terms of the three common meals -- breakfast, lunch, and dinner. That distinct three meal pattern, seems to be a modern European convention. In most societies, though, a single meal dominates the daily pattern, while the earlier meal is composed of a drink (tea, milk, coffee, etc.) and bread, gruel, or another carbohydrate source and a later meal made up of leftovers or soup. Today's "supper" is descended from the medieval European soup meal. In the rural Central American diet, for example, corn (or rice), beans, and coffee are eaten at all meals, but meat and vegetables predominate in the midday meal if a family can afford them. The evening meal may include meat or vegetables leftover from midday.

In the U.S., different meals tend to be characterized by specific foods --cold cereal or eggs and sausage for breakfast, sandwiches for lunch, etc. Pizza for breakfast would seem out of place to most consumers, though it might make no nutritional difference (Lowenberg, 1970). New foods will probably be more acceptable if they fit into the familiar meal pattern. The recent "breakfast bars" are an informative example of a popular breakfast cereal in an even more convenient form -- one which anticipated the reduced time devoted to the breakfast meal.

Meal pattern is obviously strongly influenced by lifestyle and other activities. A recent national survey in Great Britain found that only 59% ate breakfast every weekday and 22% never ate breakfast (Thomas, 1982). In the U.S. and Great Britain the midday meal lost its prominence as the main family meal along with industrialization and the urban work schedule. It was replaced by the evening meal. "Where lunch has been a snack for the housewife or eaten away from home there is much more likelihood that an evening meal, or dinner, will be the most important meal of the day. Eating a meal together is a symbolically important act in many cultures and disruption in family meal routine may be rather more disturbing to individuals than any nutritional considerations would seem to warrant." (Thomas, 1982:219). Supper, too, has become a matter of convenience rather than social congeniality. Indeed, one researcher, lamenting the "fragmentation of social relationships...within the family" proposed that "the means of re-establishing the family meal might well be worth looking into, not merely as a contribution to better nutrition, but to better family and social relationships" (Montague, 1957: 243).

The daily meal pattern in the U.S. seems to have lost some of its definition, but it does still serve as the structure by which people conceptualize foods. The evening meal tends to be the major social meal and so appears to be more central to food habits. It has been hypothesized that "people are prepared to make modifications first to those meals which are least important" (Thomas,1982: 219; also Lowenberg, 1970: 752-753). Breakfast is viewed as the "weakest link" for innovation and breakfast cereals have been cited as an example.

meal composition: Meals are also internally structured. There are, for example, the "courses" or "plates" of a formal meal; the basic elements of common fare -- drink, meat, and a carbohydrate source -- and foods that "go together" or, conversely, are considered antithetical.

The structuring of meals by courses began in the kitchens of medieval European nobility, though haphazardly organized, from our modern perspective until the 16th century (Tannahill, 1973: 222). An Italian banquet in 1570 was composed of 4 courses: "cold delicacies from the sideboard", roast meats and fowl, boiled meats and stews, and sweets from the sideboard. The British menu took shape later, in the 18th century, modelled on the French example. In British form were three courses: appetizers, soups and fish, large cuts of meat and poultry, and "afters" including vegetables and sweets (Tannahill, 1973: 335). The major purpose of the aristocratic, multi-course meals was probably neither variety in the diet nor aesthetics but rather display. Such a lavish spread, like the yam pile of a Melanesian chief (Malinowski, 1935: 82) indicated the human resources which a leader could command.

Common fare was never so elaborate. In medieval Europe it was basically bread (rye or barley), complemented when possible by something from the stockpot. Potatoes later came to substitute bread as the common staple. With industrialization and modernization, a middle-class diet emerged which is neither so elaborate as the noble cuisine nor so marginal and monotonous as the peasants' fare.

The components of the modern Western meal are generally main dish, dessert and beverage. Optional components are bread, salad, and soup. The main dish is expected to include meat and potatoes or another starch source. Other vegetables seem optional. From first course to dessert "the meal moves from savory to sweet; from a hot solid and a cold drink to a cold solid and a hot drink..." (Douglas, 1979: 17). This is an idealized scheme, and, though it may not be adhered to strictly, it does influence behavior. About half of the British housewives surveyed by Kraft Foods in 1978 "agreed that their husbands would complain if they served them a meal without meat or fish in it" (Thomas, 1982: 221).

If food habits that are peripheral are more subject to change than those that are central (Passin and Bennett, 1943; Lowenberg, 1970) then, hypothetically, the meat component of the main dish will be the most resistant to change. It might also be hypothesized that foods which honor the familiar meal structure will be more acceptable than those that do not. One of the most successful fast food chains publicizes its hamburgers as a "meat and potatoes meal". It seems unlikely that consumers will prefer to consume their major nutrients in the form of beverages or pills, for example.

References

Armenta, George
1970 *"'Puma' a High Protein Soft Drink for Guyana."* CAJAMS
3(5):289-291.

Austin, James E. and Marian R. Zeitlin (eds.)
1981 NUTRITION INTERVENTION IN DEVELOPING COUNTRIES.
Cambridge, Mass.: Oelgeschlager, Gunn and Hain, Publ., Inc.

Barker, T.C.
1978 *"Changing Patterns of Food Consumption in the United
Kingdom."* pp. 163-186 in John Yudkin (ed.) DIET OF MAN:
NEEDS AND WANTS. London: Applied Science Publishers.

Bennett, John W., Herbert Passin and H. Smith.
1943 *"Food and Culture in Southern Illinois."* AMERICAN
SOCIOLOGICAL REVIEW 7: 645-660.

Braudel, Fernand
1973 CAPITALISM AND MATERIAL LIFE 1400-1800. Miriam Kochan
(trans.), Harper & Row: New York (originally 1967).

Call, D.L.
1969 *"Economic and Marketing Factors Influencing the Potential
Acceptance of Food Substitutes."* ABSTRACTS OF PAPERS.
AMERICAN CHEMICAL SOCIETY, 158: AGFD 30.

Consumer Reports
1980 *"Vegetarian 'meats'."* CONSUMER REPORTS 45(6): 357-365.

Coon, Carleton
1951 CARAVAN. Henry Holt, New York.

Curry, P.M. and R.M. Jiobu
1980 *"Big Mac and Caneton a l'orange: Eating, Icons and
Rituals."* pp. 248-257 *in* R.B. Browne (ed.) RITUALS AND CERE-
MONIES IN POPULAR CULTURE. Bowling Green, Ohio: Bowling
Green University Popular Press.

Cuthbertson, W.F.J.
1966 *"Problems of Introducing New Foods to Developing
Countries."* FOOD TECHNOLOGY 20: 634-636.

de Garine, Igor.
1972 *"The Socio-Cultural Aspects of Nutrition."* ECOLOGY OF
FOOD AND NUTRITION 1(2): 143-163.

den Hartog, Adel
1981 *"Urbanization, food habits and nutrition: A review of situations in developing countries."* WORLD REVIEW OF NUTRITION
AND DIETETICS 38: 133-152.

Douglas, Mary
1966 *"The Abominations of Leviticus."* pp. 41-57 *in* Mary Douglas, PURITY AND DANGER: AN ANALYSIS OF CONCEPTS OF POLLUTION AND TABOO. Baltimore, Md.: Penguin.

Douglas, Mary
1979 *"Accounting for Taste.* 'PSYCHOLOGY TODAY.

Duda, Zbigniew
1974 *"Vegetable Protein Meat Extenders and Analogues."* Food and Agricultural Organization of the U.N. (AGA/MISC/74/7).

Dwyer, J.T., R.F. Kandel, L.D. Mayer, and J. Mayer.
1974 *The "New" Vegetarians. Group Affiliation and Dietary Strictures Related to Attitudes and Life Style."* JOURNAL OF THE AMERICAN DIETETIC ASSOCIATION 64(4): 376-382 (Apr).

Eppright, Ercel S.
1947 *"Factors Influencing Food Acceptance."* JOURNAL OF THE AMERICAN DIETETIC ASSOCIATION 23: 579-587.

Fallon, April and Paul Rozin
1983 *"The psychological bases of food rejection by humans."* ECOLOGY OF FOOD AND NUTRITION 13: 15-26.

Ford, Barbara
1978 *"Future Food: Alternate Protein for the Year 2000."* William Morrow and Company, New York.

Foster, George M.
1967 *"Tzintzuntzan: Mexican Peasants in a Changing World."* Boston: Little Brown.

Garb, Jane L. and Albert J. Stunkard.
1974 *"Taste Aversions in Man."* AMERICAN JOURNAL OF PSYCHIATRY 131: 1204-1207.

Garcia, John and Walter G. Hankins.
1975 "The Evolution of Bitter and the Acquisition of Toxiphobia" pp. 39-45 in Derek A. Denton and John P. Coghlan (eds.) OLFACTION AND TASTE V. N.Y.: Academic Press.

Garcia, et al.
1966 See Rozin 1978

Glazer, Nathan and Daniel Patrick Moynihan.
1963 *"Beyond the Melting Pot."* MIT Press, Cambridge.

Grivetti, L.E. and R.M. Pangborn.
1974 *"Origin of Selected Old Testament Dietary Prohibitions. An Evaluative Review."* JOURNAL OF THE AMERICAN DIETETIC ASSOCIATION 65(6): 634-638. (Dec)

Gross, Daniel and Barbara Underwood
1971 *"Technological change and caloric costs: Sisal agriculture in Northeastern Brazil."* AMERICAN ANTHROPOLIGIST, 73: 725-740.

Gupta, Santosh
1965 *"Changes in the food habits of Asian Indians in the United States: A case study."* SOCIAL SCIENCE RESEARCH 60(1): 87-99.

Harner, M.
1977 *"The Enigma of Aztec Sacrifice."* NATURAL HISTORY 86: 47-51.

Harris, L. J., J. Clay, F.J. Hargreaves, and A. Ward
1933 *"Appetite and choice of diet. The ability of the vitamin B deficient rate to discriminate between diets containing and lacking vitamin."* PROCEEDINGS OF THE ROYAL BIOLOGICAL SOCIETY, London. 113: 161-189.

Harris, Marvin
1974 *"Cows, Pigs, Wars and Witches."* NEW YORK: Random House.

Harris, Marvin and Eric B. Ross.
1978 *"How Beef Became King."* PSYCHOLOGY TODAY 12(5): 88-94.

Harvey, P.W. and P.F. Heywood.
1983 *"Twenty-five Years of Dietary Change in Simbu Province, Papua New Guinea."* ECOLOGY OF FOOD AND NUTRITION 13(1): 27-35.

Hollingsworth, Dorothy F.
1961 *"The Changing Patterns of British Food Habits Since the 1939-1945 War."* PROCEEDINGS OF THE NUTRITION SOCIETY 20:25-30.

Hopkins, Jane, John Nichols, and Leslie Chin
1983 *"Consumer acceptance of fortified weaning foods."* The case of Cerex in Guyana, Texas Agricultural Market Research and Development Center Research Report MRC 83-2, College Station, Texas.

Hull, C.L.
1943 *"Principles of Behavior."* APPLETON, Century, New York.

Itani, Jun'ichiro and A. Nishimura
1973 *"The Study of infra-human culture in Japan,"* in E.W. Menzell (ed.), PRECULTURAL PRIMATE BEHAVIOR, Basel, S. Karjer: 26-50.

Jacobs, Harry
1973 *"Taste and the Role of Experience in the Regulation of Food Intake,"* in THE CHEMICAL SENSES AND NUTRITION: 187-200.

Kariel, Herbert G.
   1966   *"A Proposed Classification of Diet."* ASSOCIATION OF
          AMERICAN GEOGRAPHERS. ANNALS 56: 68-79.

Katona-Apte, Judit.
   1975   *"Dietary Aspects of Acculturation: Meals, Feasts, and Fasts in
          a Minority Community in South Asia."* pp. 315-326 *in*
          Margaret Louise Arnott (ed.) GASTRONOMY: THE
          ANTHROPOLOGY OF FOOD AND FOOD HABITS. Paris, The
          Hague: Mouton Publishers. World Anthropology.

Kaunitz, Hans.
   1956   *"Causes and Consequences of Salt Consumption."* NATURE
          178: 1141-1144.

Keegan, Warren
   1980   *"Multinational Marketing Management."*

Kent, John
   1977   *"How Ersatz Foods Shortchange You,"* SCIENCE DIGEST: 20.

Knutsson, Karl Eric and Ruth Selinus.
   1970   *"Fasting in Ethiopia. An Anthropological and Nutritional
          Study."* AJCN 23: 956-969.

Kolar, C.W., I.C. Cho, and W.L. Watrous
   1979   *"Vegetable protein applications in yogurt, coffee creamer
          and whip toppings."* JOURNAL OF THE AMERICAN OIL
          CHEMISTS' SPCOETU. 56(3): 389-391.

Korff, S.1.
   1966   *"The Jewish dietary code."* FOOD TECHNOLOGY 20:926-??.

Kracht, Uwe
   1972   *"The Economics and marketing of protein-rich food
          products."* Status Report: Incaparina and Competitors. *in*
          P.L. White and N. Selvey (eds.) Proceedings of the
          Western Hemisphere Nutrition Congress III Futura Publishing
          Co., Mt. Kisco, New York: 99-102.

Leach, Edmund
   1964   *"Anthropological aspects of language: animal categories
          and verbal abuse."* *in* E.H. Lenneberg (ed.) NEW DIRECTIONS
          IN THE STUDY OF LANGUAGE. Cambridge, Massachusetts: MIT
          Press.

Lowenberg, Miriam E.
   1970   *"Socio-cultural Basis of Food Habits."* FOOD TECHNOLOGY
          24(7): 27-32.

Lusas, E.W.
1979 *"Food Uses of Peanut Protein."* JOURNAL OF THE AMERICAN
OIL CHEMISTS' SOCIETY. 59(3): 425-30.

Malinowski, Bronislaw
1935 CORAL GARDENS AND THEIR MAGIC, VOLUME 1: SOIL-TILLING
AND AGRICULTURAL RITES IN THE TROBRIAND ISLANDS.
Bloomington: Indiana University Press.

Martinez, Wilda
1979 *"Functionality of Vegetable Proteins Other than Soy."*
JOURNAL OF THE AMERICAN OIL CHEMISTS' SOCIETY
56(3): 280-284.

McCann, Mary B. and Martha F. Trulson.
1957 *"Our Changing Diet."* JOURNAL OF THE AMERICAN DIETETIC
ASSOCIATION 33: 358-365.

McKenzie, John C.
1967 *"Social and Economic Implications of Minority Food Habits."*
PROCEEDINGS OF THE NUTRITION SOCIETY 26: 197-205.

Miner, B.D. and W.W. Gallimore
1977 *"Soy Protein Use can increase 71 Percent by 1985."* U.S.
DEPARTMENT OF AGRICULTURE, Farmer Cooperative Reprint 4.

Montague, Montague Rhodes Ashley.
1957 *"Nature, Nurture and Nutrition."* AJCN 5: 237-244.

Nalbandian, A., J.G. Bergan, and P.T. Brown.
1981 *"Three Generations of Armenians: Food Habits and Dietary
Status."* JOURNAL OF THE AMERICAN DIETETIC ASSOCIATION
79(6): 694-699.

New, Peter Kong-Ming and Rhea Pendergrass Priest.
1967 *"Food and Thought: A Sociologic Study of Food Cultists."*
JOURNAL OF THE AMERICAN DIETETIC ASSOCIATION 51: 13-18.

Nicholls, Charles
197? *"The future of non-traditional protein foods. ms for the U.S.
Food and Drug Administration."*

Ogbeide, 0.
1974 *"Nutritional Hazards of Food Taboos and Preferences in Mid-
West Nigeria."* AMERICAN JOURNAL OF CLINICAL NUTRITION
27(2): 213-216. (Feb)

Ortiz, Sutti R. de
1973 UNCERTAINTIES IN PEASANT FARMING: A COLOMBIAN CASE.
London: The Athlone Press, University of London and New
York: Humanties Press Inc.

Page, Louise and Bertha Friend.
1978 *"The Changing United States Diet."* BIOSCIENCE 28: 192-197.

Passin, Herbert and John W. Bennett.
1943 *"Social Processes and Dietary Change,"* pp. 113-123 in THE
PROBLEM OF CHANGING FOOD HABITS. 1941-1943.
NATIONAL RESEARCH COUNCIL. BULLETIN No. 108.

Peryam, David R.
1963 *"The Acceptance of Novel Foods."* FOOD TECHNOLOGY

Phillips-Conroy, Jane
1983 untitled ms.

Pyke, Magnus.
1975 *"The Influence of American Foods and Food Technology in
Europe"* pp. 83-95 in C.W.E. Bigsby SUPERCULTURE. Bowling
Green, Ohio: Bowling Green University Popular Press.

Pyke, Magnus.
1978 *"The Evolution of Animal Protein in the Human Diet"* pp. 47-71
in Aaron M. Altschul and Harold L. Wilcke (eds) NEW
PROTEIN FOODS. VOL. 3. ANIMAL PROTEIN SUPPLIES, PART A.
FOOD SCIENCE AND TECHNOLOGY: A SERIES OF
MONOGRAPHS. N.Y., San Francisco, London: Academic Press.

Rapport, Roy
1968 *"Pigs for the ancestors: Ritual in the ecology of a New
Guinea people."* New Haven, Yale University Press.

Richter, C.P.
1939 *"Salt-taste thresholds of normal and adrenalectomized
rats."* ENDOCRINOLOGY 24: 367-371.

Richter, C.P.
1942 *"Increased dextrose appetite of normal rats treated with
insulin."* AMERICAN JOURNAL OF PHYSIOLOGY, 135: 781-787.

Rozin, Elisabeth.
1982 *"The Structure of Cuisine,"* pp. 189-203 (Chap. 11) in Lewis
M. Barker (ed) THE PSYCHOBIOLOGY OF HUMAN FOOD
SELECTION. Westport, Conn. Avi Publishing Co., Inc.

Rozin, Paul.
1976 *"The Selection of Foods by Rats, Humans, and Other
Animals."* ADVANCES IN THE STUDY OF BEHAVIOR 6: 21-76.

Sanford, D.
  1968 *"Unfoods: Do You Know What You're Eating?"* NEW REPUBLIC
        158: 13-15. (May 18)

Schnakenberg, David
  1983 *"Use of Soy Proteins in Military Feeding Systems presented to
        the Second Meeting of NATO Panel VIII, Research Study
        Group 8, Nutritional Aspects of Military Feeding Systems."*
        Zeist, Netherlands. October 17-21, 1983.

Schmitt, Martin
  1952 *"Meat's Meat": An account of the flesh-eating habits of
        western Americans.* WESTERN FOLKLORE, 11: 185-203.

Schwerin, H.S., J.L. Stanton, A.M. Riley Jr., and B.E. Brett
  1981 *"How Have the Quantity and Quality of the American Diet
        Changed During the Past Decade?"* FOOD TECHNOLOGY
        35(9): 50-57.

Scrimshaw, N.S.
  1980 *"A Look at the Incaparina Experience in Guatemala. The
        Background and History of Incaparina."* FOOD AND
        NUTRITION BULLETIN, 2(2): 1-2.

Sellers, Stephen
  1984 *"Diet patterns and nutritional intake in a Costa Rican com-
        munity."* ECOLOGY OF FOOD AND NUTRITION, 14: 205-218.

Senti, F.R.
  1974 *"Soy Protein Foods in U.S. Assistance Programs,"*

Shack, Dorothy
  1975 *"Taster's Choice: Social and Cultural Determinants of Food
        Preferences." in* John Yudkin (*ed.*) DIET OF MAN: NEED AND
        WANTS. London: Applied Science Publishers: 209-224.

Sharon, Irving
  1965 *"Sensory properties of food and their function during
        feeding."* FOOD TECHNOLOGY, January: 35-36.

Shawn R.L.
  1972 *"Incaparina: The Market Development of a Protein Food."*
        TROPICAL SCIENCE 14(4): 347-371.

Shaw, R.L.
  1975 *"Incaparina: A low cost protein-rich food product."* The PAG
        Compendium: The Collected Papers Issued by the Protein-
        Calorie Advisory Group of the United Nations 1956-1973.
        World Press, John Wiley Sons, New York. Volume G: 118-125.

Shurtleff, William
 1980 *"Dr. Henry Miller: Taking Soymilk Around the World,"*
 SOYFOODS: 28-36.

Simoons, Frederick John.
 1978 *"Traditional Use and Avoidance of Foods of Animal Origin: A Culture Historical Overview."* BIOSCIENCE 28: 178-184.

Stoneman, Zolinda and Gene H. Brody.
 1981 *"Peers as Mediators of Television Food Advertisements Aimed at Children."* DEVELOPMENTAL PSYCHOLOGY 17(6): 853-858.

Tambiah, S.J.
 1969 *"Animals are Good to Think and Good to Prohibit."* ETHNOLOGY 8: 423-459.

Tannahill, Reay
 1973 *"Food in History."* STEIN AND DAY: New York.

Thomas, J.E.
 1982 *"Food Habits of the Majority: Evolution of the Current UK Pattern."* PROC. NUTR. SOC. (England) 41(2): 211-228.

Todhunter, E.N.
 1973 *"Food Habits, Food Faddism and Nutrition."* WORLD REVIEW OF NUTRITION AND DIETETICS 16: 286-317.

Welsh, Thomas
 1979 *"Meat and dairy analogs from vegetable protein."* JOURNAL OF THE AMERICAN OIL CHEMISTS' SOCIETY 56(3): 404-406.

Wenknam, Nao S. and Robert J. Wolff
 1970 *"A Half-Century of Changing Food Habits among Japanese in Hawaii."* JOURNAL OF THE AMERICAN DIETETIC ASSOCIATION 57: 29-32.

Williams, Sue Rodwell.
 1973 *"Cultural, Social and Psychological Influences on Food Habits."* (chap. 13) *in* NUTRITION AND DIET THERAPY. Saint Louis: C.Y. Mosby.

Wilson, A.T.M.
 1962 *"Nutritional Change: Some Comments from Social Research."* PROCEEDINGS OF THE NUTRITION SOCIETY, 20: 133-137.

Wilson, Christine Shearer.
 1975 *"Rice, Fish, and Coconuts -- The Bases of Southeast Asian Flavors."* FOOD TECHNOLOGY 29(6): 42-44.

Winikoff, Beverly
 1978 *"Changing Public Diet."* HUMAN NATURE 1(1): 60-65.

Winninydji and A.R. Peile
1978 *"A desert Aborigine's view of health and nutrition."* JOURNAL OF ANTHROPOLOGICAL RESEARCH 34(4): 497-523.

Wise, R.P.
1980 *"The Case of Incaparina in Guatemala."* FOOD AND NUTRITION BULLETIN 2(2): 3-8.

Wolf, Walter
1976 *"Production Estimates and Market Outlets." in* EDIBLE SOY PROTEIN PRODUCTS.

Yalman, Nur.
1969 *"On the Meaning of Food Offerings in Ceylon."* pp. 81-96 *in* Robert F. Spencer (eds.) FORMS OF SYMBOLIC ACTION. Seattle: University of Washington Press.

Young, Paul Thomas
1949 *"Food-seeking drive, affective process, and learning."* PSYCHOLOGICAL REVIEW 56: 98-121.

Yudkin, John.
1956 *"Man's Choice of Food."* LANCET 270: 645-649. page 117